PH.

MW01487365

SCIENCE,

AND THE

$\mathscr{S}$OVEREIGNTY

OF $\mathscr{G}$OD

Other P&R Works by the Author

God-Centered Biblical Interpretation

Symphonic Theology: The Validity of Multiple Perspectives in Theology

The Returning King: A Guide to the Book of Revelation

The Shadow of Christ in the Law of Moses

Understanding Dispensationalists

What Are Spiritual Gifts?

PHILOSOPHY,

SCIENCE,

AND THE

$\mathcal{S}$OVEREIGNTY OF $\mathcal{G}$OD

VERN S. POYTHRESS

P&R PUBLISHING

P.O.BOX 817 • PHILLIPSBURG • NEW JERSEY 08865-0817

Printed in the United States of America

Library of Congress Cataloging-in-Publication Data

Poythress, Vern S.
   Philosophy, science, and the sovereignty of God / Vern S. Poythress.
      p. cm.
   Originally published: Nutley, NJ : P&R Pub., 1976.
   Includes bibliographical references.
   ISBN-10: 1-59638-002-0 (pbk.)
   ISBN-13: 978-1-59638-002-8 (pbk.)
   1. Bible and science. 2. Science—Philosophy. 3. Christianity—Philosophy. I. Title.

BS650.P69 2004
261.5'5—dc22

2004057517

ACKNOWLEDGEMENTS

This book is the fruit of three years' study at Westminster Theological Seminary in Philadelphia. During that time, I learned much from a company of scholars whom I respect. As a result, it is frequently no longer possible for me to identify what is mine and what is my inheritance from others. In some ways I feel that this book is little more than a patchwork of other people's ideas. Nevertheless, I am responsible for the whole that has emerged. Whatever weaknesses there may be in that whole are my fault.

I must give special thanks to Prof. Cornelius Van Til for his presuppositional apologetic stance and his striving after radical critique of unbelief; to Profs. Edmund P. Clowney, Richard B. Gaffin, and Harvie M. Conn for their instruction in biblical theology; to Profs. Meredith G. Kline, O. Palmer Robertson, and Harvie M. Conn for their material on covenants; to Prof. John M. Frame for his teaching on Scripture and language; to Prof. Kenneth L. Pike for specific tools of linguistic analysis and philosophical framework-making.

Thanks are due to John Frame for his patience in reading and criticizing several successive drafts of the manuscript.

I should be remiss if I did not also acknowledge my profound debt to members of the cosmonomic school of philosophy, and especially to Herman Dooyeweerd. Radically as my own stance differs from theirs, I have received much useful stimulus from their writings.

Above all, may the Lord himself be praised, who has given me the life and strength and insight necessary to finish this work.

All Bible quotations except those explicitly marked are from the Revised Standard Version.

TABLE OF CONTENTS

v

LIST OF TABLES

INTRODUCTION

Few evangelicals need to be convinced that it is important for Christians to say something coherent about modern science. A spate of evangelical books on the Bible and science testify to the continuing need. For one thing, the educated secularist regards the battle between the Bible and Darwinian evolution as over—and he thinks that evolution has won. So the evangelical press methodically turns out books about evolution, to undermine that easy assumption.

In the twentieth century, however, Darwinian evolution is no longer the unique focus of controversy. Far more powerful than evolution itself is an atmosphere, an atmosphere in which Rudolf Bultmann can make his famous statement that "it is impossible to use electric light and the wireless and to avail ourselves of modern medical and surgical discoveries, and at the same time to believe in the New Testament world of spirits and miracles."[1] We live in an atmosphere in which the liberal Christian feels that intellectual integrity demands his giving up many elements in the biblical story. He may even feel religiously and emotionally attracted to miracles, but he "cannot" accept that they happened. He may feel that there is something unstable and subjective about modern destructive biblical criticism, but he is told that this criticism is the most advanced "scientific" tool that we have.

We breathe an atmosphere, in fact, in which not only evolution, but engineering, psychology, medicine, sociology, linguistics, anthropology, historiography, archaeology, art, music, and philosophy are all summoned to the task of undermining biblical teaching. And the "atmospheric" quality of their effects, more than any specific argument, makes their position all the more effective because all the more subtle and irresistible.

It is not my purpose to respond directly to all of this. Francis Schaeffer does it, evangelical answers to the liberals do it, apologists do it. In fact, part of the problem may be that too often evangelicals have been content *just* to respond. The problems are posed by the liberals, and evangelicals react with answers. The problems are posed by science, and evangelicals react with answers. No doubt this has value. We should praise God for the way that he has used it. But mere reaction has weaknesses. The problems come to evangelicals already in unbiblical terms, because the problems are posed by the *secular* culture. Too often the answers have been patchwork. Too often the answers have been still partly caught in a non-Christian problematic, and so have lacked conviction. (For example, the liberal dynamistic view of revelation has sometimes provoked a fundamentalist static view of revelation, with little appreciation for the development from Old Testament to New Testament. Liberal vaunting of science has produced fundamentalist rejection of science.)

Hence I wish to concentrate in this book on the positive task of uncovering some biblical foundations for science and the philosophy of science. Most of what I say is more an introduction to philosophy of science than a treatment of special problems in philosophy of science. The question of basic orientation is at stake.

There are problems in covering such a broad field. Constructing a framework for doing science involves, eventually, saying something about everything that there is. One must speak in generalities. But if one becomes too general, he becomes trite or obscure. If one becomes too specific, he is likely to lose sight of the forest for the trees. I have endeavored to compromise. To facilitate the compromise, two special devices have been introduced into the text: (1) a detailed numerical system of outlining, and (2) technical terminology. Neither of these devices is strictly necessary. But without them, this book would have grown to unmanageable length.

Numerical section numbers have been used to divide the text into successively smaller units. For example, chapter 2 on "ontology" is divided into subsections 2.1, 2.2, 2.3, and 2.4. Section 2.4 on "Creation" is in turn subdivided into sections 2.41, 2.42, and 2.43. And so on. It is best, I think, for a reader to ignore this numbering at first, until he has grasped some of the detail. At a later stage (particularly when section 3.35 is understood), the numbering will help to show how the topics are connected, and to show my justification for treating topics in the order and with the emphasis that I have used. At a later stage, in other words, the numbering system can help one to see the generalities in addition to detail.

The second device used is technical terminology. Technical terms are introduced one by one in the text, and are thereafter capitalized to distinguish them from the words of ordinary English. In addition, a glossary has been provided at the back of the book to summarize the meanings of the terms. However, the technical terms themselves have a good deal of vagueness and imprecision about them. You must not suppose that a technical term has a perfectly precise sense, exactly the same sense every time that it is used. The technical terms are essentially like new words in English vocabulary (indeed, some of them *are* newly coined). I use the word "description" rather than "definition" in introducing new terms, to remind readers that my "definitions" should be read sympathetically and not pressed for mathematical precision.

Once again, this device can be largely ignored at first; many of the technical terms have a meaning close enough to ordinary English to allow the reader reasonable progress even when he ignores distinctions. Moreover, a large number of terms are introduced simply to describe the study of various fields. For example, Theology Proper is the study of God, Aesthetics is the study of the Aesthetic Function, Ktismatology is the study of Creation, and so on. None of these special terms for "studies" need be mastered; the main point is that almost any item of interest can be made the subject-matter for human investigation. At a later stage, the reader will find the technical terms more important, because they serve as pegs or frameworks by which modern philosophy and science can be more easily compared to biblical teaching.

NOTES TO INTRODUCTION

1. Rudolf Bultmann, "New Testament and Mythology," *Kerygma and Myth; a Theological Debate*, ed. Hans Werner Bartsch (London: S.P.C.K., 1957), p. 5. Bultmann comments further, "The various impressions and

speculations which influence credulous people here and there are of little importance, nor does it matter to what extent cheap slogans have spread an atmosphere inimical to science. What matters is the world view which men imbibe from their environment, and it is science which determines that view of the world through the school, the press, the wireless, the cinema, and all the other fruits of technical progress" (*ibid.*, n. 1). Of course, Bultmann is concerned not so much with the question whether the secularist's "world view," is *true*, but with the question of how we *communicate* to secularists. Nevertheless, because he thinks that a direct challenge to this world view is wrong, he emasculates the gospel in trying to communicate it.

Chapter 1

ORIENTATION

The word 'science' occurs only twice in the King James Version, namely in Daniel 1:4 and I Timothy 6:20. Both times it means simply "knowledge," not "science" in our twentieth-century sense. Modern versions like the Revised Standard Version, the New English Bible, and the New American Standard Version therefore use 'knowledge' or the equivalent. Does this mean that the Bible says nothing relevant to modern science? Hardly. But it means that understanding the Bible's bearing on science is more difficult.

The task is difficult partly because it is hard to know what in the Bible to appeal to. Each person wants to find in the Bible what agrees with his own preconceptions, his own life-style, his own values. No one can come to the Bible with his mind a "blank slate." He at least has to know how to read, or how to understand the language in which someone else reads to him. Furthermore, everyone comes with a basic orientation either of trusting what the Bible says because it is God's word, or of distrusting it. Everyone has some vague idea of what he is likely to find there.

Is this bad? Simply to *have* preconceptions and life-style and values is not bad. Everything depends on what they are. So let me say what is *my* way of approaching the Bible and discussing the relation of the Bible and science. Others may not agree with me, but at least they will know how I am going to proceed. If they do not agree at the beginning, they may still come to agree later on. No one need be discouraged!

I will discuss (1.1) my presuppositions, (1.2) what tools and in-

1

sights I bring to the Bible and to science, and (1.3) what is my purpose.

1.1 *Presuppositions*

By 'presupposition' I mean a belief or disposition to which one clings for life and death, and which one does not allow to be refuted by evidence. Let me illustrate with a hypothetical case. Suppose that Lydia is a believer in Christ. Lydia's fundamentalist pastor stands up in the pulpit and announces that on the basis of the latest archaeological discovery in Palestine, it is no longer possible to believe that Christ rose from the dead. Her pastor then resigns his pastorate. What does Lydia do? She may want to find out more about this supposed "discovery." But she continues to believe in Christ. She trusts in Christ more than in her pastor, more than in the judgments of archaeologists. She "presupposes" that Christ did rise.

Or suppose that Joan is an unbeliever. Lazarus returns from the dead, appears to Joan, and warns her that if she does not repent, she will go to hell. Even so, she continues in her unbelief, according to Luke 16:27-31—unless God is merciful to her and changes her heart (Ezek. 36:25-27). She "presupposes" that Christianity is not and cannot be true.

Now, my own presupposition is that Yahweh is who he is. It is unthinkable that Yahweh should be other than who he is. Hence it is proper that this should be a firm basis for everything that I do, including what I say in this book.

I must explain something of what I mean. In the first place, when I speak of Yahweh, I mean the God who has told us about himself in the Bible, which is his word. I am not talking about some vague, general "theism." No doubt the word 'God' is often used by people in cases where they have no intention of identifying "God" with the God of the Bible. In using the word 'God,' they are not talking about Yahweh. Hence they are simply building in their minds a hypothetical god. We ought not to be fooled by the fact that they still use the word 'God.' For the sake of clarity, I will use 'God' when I am speaking of Yahweh and 'god' or 'idol' when I am

not like Yahweh and does not do
ıg in the Bible.
to make the point that Cornelius
ınly theism worthy of the name is
, Trinitarian theism.[1] Jesus says,
he Father, and no one knows the
ıne to whom the Son chooses to
n the way, and the truth, and the
but by me" (John 14:6). Hence
: knows God through Jesus Christ
ıne who goes ahead and does not
not have God" (II John 9).

__, that "Yahweh is who he is," I mı ... ı summarize what the Bible says, not to go off on a speculative tack. I think first of all of the fact that Jesus Christ is Yahweh (I Cor. 12:3; Acts 10:36; Heb. 1:10-12). He is my Lord, to whom I owe unconditional allegiance, and to whom I·am to entrust my life and my salvation. Thus, instead of saying that I presuppose that Yahweh is who he is, I could equally well have explained my presuppositions in the words of Van Til:

> As Christians we are not, of ourselves, better or wiser than were the Pharisees. Christ has, by his word and by his Spirit, identified himself with us and thereby, at the same time, told us who and what we are. As a Christian I believe first of all in the testimony that Jesus gives of himself and his work. He says he was sent into the world to save his people from their sins. Jesus asks me to do what he asked the Pharisees to do, namely, read the Scriptures in light of this testimony about himself. He has sent his Spirit to dwell in my heart so that I might believe and therefore understand all things to be what he says they are. I have by his Spirit learned to understand something of what Jesus meant when he said: *I am the Way, the Truth, and the Life.* I have learned something of what it means to make every thought captive to the obedience of Christ, being converted anew every day to the realization that I understand no fact aright unless I see it in its proper relation to Christ as Creator-Redeemer of me and my world. I seek his kingdom and its righteousness above all things else.[2]

In the third place, the fact that Jesus Christ is *Lord* implies that

I ought to heed and to take to heart everything that he says. And, as Van Til says, "He has written me a letter."[3] Because the Bible is the word of my *Lord,* I try to give heed to everything that it says with the kind of obedience that my Lord deserves.

I could wish that this were all that I needed to say about the status of the Bible. But, unfortunately, Bible-believing Christians *do* disagree among themselves to some extent about what the Bible teaches. Some of these differences have a great influence on the development of Christian philosophy. So I will specify more exactly: my own interpretation of the Bible is like that of the Reformation. More specifically still, it is like that of the Westminster Confession of Faith.[4] I cannot take space here to argue about it. I am aiming at Reformed philosophy. This does not necessarily mean that, if others do not like the Reformation, they will not follow me. I simply want to be frank about my own biases.

Though these are my presuppositions, I am not saying that I decided on these presuppositions by an arbitrary, sudden "leap of faith." Actually, I cannot trace exactly how I came to where I am. All I am saying is that these are in fact my sure basis for doing philosophy, and that they *ought* to be other people's basis.

Isn't there a problem in the fact that I have a bias? I think not. Positively, the Bible indicates that people ought to approach God's world with this kind of bias. Negatively, the Bible indicates that an unbeliever also has a bias, and a bad bias at that. He is a covenant-breaker, a rebel against God.

Cornelius Van Til has already said much about this, so I will not dwell on it. I should only like to make one point. Perhaps the easiest way for a believer to illustrate that the unbeliever has a bias is to confront him with the believing attitude that I have sketched out above. Then the believer says to the unbeliever, "You too ought to look at the world in a Christian way." To this the unbeliever could respond in three basic ways. (1) He could become a Christian, in which case he would begin to have (though imperfectly) a Christian bias. (2) He could say, "I have a religious bias too. I'm against Christianity as you describe it and for Buddhism (or atheism, materialism, etc.)." (3) He could say, "It's bad for you to start with any bias. You must clear away the 'slate' and try to approach the world fresh, with no biases at all. That is what I try to do."

Now consider the dialog that might follow.

Christian: "I see that you think that it is all right to do your thinking without Christian biases. Now I would be the first to affirm that extraordinary feats of thinking and remarkable insights have been achieved by people who are not Christians. That's not the question. The Christian faith says that people *ought* to approach the world with Christian bias. Notice the ethical force there. You evidently disagree with that 'ought.' Hence you have already rejected the Christian faith (not that you will necessarily reject it forever, but you are rejecting it right now). You have a religious bias."

The unbeliever: "I haven't made a religious commitment at all; I've simply kept myself open for various possibilities."

Chr.: "You *are* denying, by action if not word, that you have the clear *obligation* to think with Christian bias."

Unbel.: "No, I'm keeping myself open."

Chr. "Is that 'openness' better than Christian bias?"

Unbel.: "I don't know." (If he said yes, he would clearly be guilty of anti-Christian bias.)

Chr.: "Ought you to be open?"

Unbel.: "I don't know."

Chr.: "You're rebelling against God insofar as you don't listen to him."

Unbel.: "O.K., I *do* think that I ought to be open until I can really get convinced that Christianity is true."

Chr.: "Your bias is in that 'ought' and in the fact that you won't come to Christ now."

Unbel.: "We're quibbling over a term. You are in effect rejecting the Christian faith, and that will color your thinking inasmuch as you won't use the Bible as an unimpeachable authority. This is a religious bias against Christianity."

So perishes the myth of the autonomy or neutrality of thought. Thinking and discussion is not done in a "vacuum," but by people who have certain attitudes toward God's claims in the Bible.

1.2 *In medias res*

So far I have talked about my presuppositions, about what is "nonnegotiable" for me. But presuppositions are not the only thing that we bring to the task of doing Christian philosophy of science. Everyone has the background of his personal history, his knowledge of people, linguistic tools and historical tools for understanding the Bible better, and so forth. I do not intend to shove these things aside either, as if I could start fresh like Descartes. My personal baggage is one of God's gifts to me. But I must be careful. "Personal baggage," unlike the Bible, is fallible.

I do not intend to make a *sharp* distinction between what is nonnegotiable (presuppositions) and what is negotiable. About certain of the Bible's teachings I am only relatively sure, so these are only relatively nonnegotiable. This is another way of saying that Christian growth is a process, including growth in what we know as well as in what we do.

Moreover, I hope that I am making demands on others similar to what I am making on myself. I do not expect readers to forget their present "knowledge," but to shake it up in the light of Scripture, to rearrange their world view, to repudiate what they see is un-scriptural.

What I am saying about my "personal baggage" may seem trivial, but I think that it is worth saying. Certain writings by Christians in our day have made it a point to strip themselves down to some few basic truths before proceeding to build a larger system.[5] Gordon H. Clark appeals to the law of contradiction to decide among religious world views.[6] John W. Montgomery appeals to historical evidence for the resurrection.[7] I will not dispute that there is methodological value in seeing what conclusions follow from limited assumptions; and various sets of starting assumptions are interesting. However, the judgments about what does and does not follow from the assumptions in question are themselves influenced by the judge's "personal baggage" (see Appendix 4). There is nothing embarrassing about that. It is part of being man, the creature of God, depending on God for knowledge of the truth.

1.3 *Problems of philosophy of science*

Just what kind of questions does philosophy of science deal with? Well, there are many such questions, but I propose to focus on three. Scientific activity generally presupposes, within a scientific community, some kind of answers, vague or specific, to three interconnected basic problems: (a) what are we studying; (b) how do we come to know what we know about it (scientific method); and (c) what is the value of this study. In a word, a science relies on (a) ontology, (b) an epistemology or, more generally, a methodolgy, and (c) an axiology or system of values. Part (c) includes both justification for choosing one special problem over another, and means of evaluating the quality and validity of scientific achievements. These three areas will be the subjects for discussion in the next three chapters.

But there is a danger here. The danger is that we will define science and formulate expectations about science too much in terms of the science that we see in the twentieth century. The particular form that sciences have taken in our time is greatly influenced by a historical development that has contained both good and bad influences. The existing form of sciences therefore cannot serve as a norm for us.

Hence I propose, before "homing in" on twentieth-century science, to consider the three basic problems in a much more general setting. How do we answer, from a Christian point of view, the following questions: (a) what is there? (ontology); (b) how does everything function? (methodology); and (c) why is it there? (axiology). All three of these questions are patently metaphysical questions. That does not mean that we are obliged to give "metaphysical" answers in the traditional sense. *Some* kind of answer is nevertheless needed for the philosophy of science.

NOTES TO CHAPTER 1

1. Cornelius Van Til, *The Defense of the Faith* (Philadelphia: Presbyterian and Reformed, 1955), pp. 9-13, 114f.; *idem, Apologetics* (unpublished syllabus; Chestnut Hill, Pa.: Westminster Theological Seminary, 1959), p. 66; and many places elsewhere.
2. Cornelius Van Til, "My Credo," *Jerusalem and Athens,* ed. E. R. Geehan (Philadelphia: Presbyterian and Reformed, 1971), pp. 4-5.

3. *Ibid.*, p. *5*.

4. Of course—need I say this?—I do not mean to sanction every turn of phrase in the Westminster Confession. Some phrases in it, like "light of nature" and "covenant of works," are wide open to misunderstanding.

5. Herman Dooyeweerd appears to speak of such a "stripping down" in *A New Critique of Theoretical Thought* (Philadelphia: Presbyterian and Reformed, 1969), II, 73-74. In his case, however, he would not describe what remains as "basic truths," but rather as a religious direction.

6. Gordon H. Clark, *A Christian View of Men and Things* (Grand Rapids: Eerdmans, 1952), pp. 30ff.

7. John W. Montgomery, *Where Is History Going? Essays in Support of the Historical Truth of the Christian Religion* (Grand Rapids: Zondervan, 1969); *idem, The Shape of the Past: An Introduction to Philosophical Historiography* (Ann Arbor: Edwards Brothers, 1962).

Chapter 2

ONTOLOGY

First, consider the question, "What is there?" An enormous variety of answers have been given to this question in the history of philosophy. That itself is one indication that the question is not a clear one. We feel perplexed as to just what kind of answer the inquirer desires, and what kind of answer would effectively meet his needs. Suppose some one asks, "What is there in the cupboard?" or "What is there on television tonight?" or "What is there to account for his moroseness?" Well, we may not know the answer, but we have a feeling that we know what it would be like to give an answer. On the other hand, the question "What is there?" seems to be asking "What is there *in general?*," and we are at a loss where to begin. One could give a facetious answer that would be true enough: "Everything," or "blueberry pancakes, elm leaves, eggbeaters, and so on." But if the question is a serious question, the best course is to ask what kind of concern generated the question in the first place.

One possible reason for asking the question is religious distress. A person may become disoriented and alienated in the world, as a result of rebellion against God. Out of his distress he cries out for some kind of basis for meaning in life: "What is there?" I will in fact treat the question as a religious question and give it a religious answer. I do not claim that my answer is the only correct way of answering the question. I claim that it is *a* biblical answer, but obviously I do not include every teaching from the Bible that I could include. I include, from the Bible's teaching, what may be helpful in giving a person some general orientation and structure. I aim at a Refined answer, in the sense to be described later (5.32).

9

2.1 Creator/creation

What is there? There is God, and then there is his creation. Everything except God is a creation of God. This is a most important truth to reckon with, if only because God thought it fitting to stress this in the very first part of his inscripturated word to Israel. "In the beginning God created the heavens and the earth" (Gen. 1:1). "Thus the heavens and the earth were finished, and all the host of them" (Gen. 2:1). "These are the generations of the heavens and the earth when they were created, in the day that the Lord God made the earth and the heavens" (Gen. 2:4; my translation). And elsewhere in Scripture: "For in six days the Lord made heaven and earth, the sea, and all that is in them" (Exod. 20:11; cf. Neh. 9:6; Col. 1:16).[1]

In particular, this eliminates the picture that would make "law" a *tertium quid,* an intermediate "being" between God and creation.[2] H. Evan Runner is wrong in saying that "the law, which is the boundary between God and cosmos, is neither the divine being nor is it created. It is, with God and cosmos, a third mode of being. God *creates* the cosmos, *puts* the law."[3] Prof. Runner cannot sustain this view from Scripture. God's law is divine (cf., e.g., Ps. 119:89), and when it becomes inscripturated in the Bible, it is also creaturely (it is in human language and written on stone or other writing material). I must defer a fuller discussion of law to 3.324.

The maintenance of the distinction between God and creation is important because God only is to be worshiped (Deut. 6:13; Matt. 4:10; Rev. 22:9). God is the Lord and his creatures are his servants (Ps. 119:91). God always was (he is eternal), whereas creation had a beginning (John 1:1). Any one of these characteristics of God (deserving of worship, lordship, eternality) could serve to describe the difference between God and creatures. So I suggest the following.

Description. *Creation* is everything that has been created by God, i.e., everything that has a beginning.
Description. A *Creature* is a thing in Creation.

We can also speak of the study of these things.

Description. *Theology Proper* is the study of God.

The word 'theology' has sometimes been used in the past to denote Theology Proper.[4] But I reserve the term 'theology' for another use (6:123).

Description. *Ktismatology* is the study of Creation.

Theology Proper and Ktismatology are not two rigidly separated, nonoverlapping disciplines. The boundaries of the two disciplines are not clearly defined. One can hardly say much about God without talking about his relations to the Creation, and neither can one say anything about Creation without implying something about God (namely, that he ordained it so to be). Thus the two studies flow into one another, and it becomes a matter of emphasis which study a person is doing. One can, as it were, say, *"God* made the rose" or "God made the *rose"*— in the first case Theology Proper, in the second case Ktismatology. On the other hand, there is no mingling of God and Creation into something neither divine nor created.

2.2 *The Creator*

I do not intend to dwell long on Theology Proper, because systematic theology has already said so much. But I should point out that we can make a further ontological distinction within God. God is three persons, Father, Son, and Holy Spirit. These are one God in a way that we shall never fully comprehend. As a result, we can describe corresponding studies.

Description. *Patrology, Christology, and Pneumatology* are, respectively, the study of the Father, the Son, and the Holy Spirit.

If we were unable really to separate between Theology Proper and Ktismatology, much less shall we be able to separate Patrology, Christology, and Pneumatology. Here we have what could at most be a difference of emphasis.

2.3 *The Mediator*

Description. The *Mediator* is "the man Christ Jesus" (I Tim. 2:5).

Description. *Incarnate Christology* is the study of the Mediator Jesus Christ.

This, of course, is not meant to describe a "third realm of existence" besides God and Creation, but to deal with the occurrence of someone who is *both* God *and* Creature. As far as I am concerned, "Incarnate Christology" could have remained as a not-explicitly-defined subdivision of Christology (2.2), but it seemed to me best to guard against any tendency to minimize the full humanity and Creatureliness of Jesus Christ, or to minimize the marvelous and praise-evoking newness of the Incarnation.

Incidentally, by distinguishing between "Christology" (2.2) and "Incarnate Christology" (2.3) I do not in the least intend to suggest that the "real" Son is a mysterious unknown being lurking behind the Incarnation. No, Incarnate Christology is simply the study of the Son after he became man. Neither am I saying that we could, after the fall, come to know the Son if it had not been for the Incarnation. I *am* saying, "Remaining what he was, he became what he was not." Christology is a larger subject than Incarnate Christology because the Son was active in creation and in the Old Testament.[5] Of course, this leaves open the mystery of Old Testament theophanies. These are undoubtedly a foreshadowing and prolepsis of the Incarnation, yet Christ has not yet become man. I am inclined to include the theophanies under Incarnate Christology, but the exact boundaries of study are not important.

One may now ask, "What has happened to Scripture?" Is the Scripture God or Creation or both? Scripture is both divine (has divine authority, holiness, infallibility, etc.) and human (it is human language). It is personal, but not a person. It is creaturely *language,* but not a Creature. That is, to anticipate 2.4, it is not an angel, a man, an animal, a plant, or a nonliving thing ("mineral").

Now, to be sure, God's word is written on stone, or papyrus, or parchment, or paper. But it would be slightly misleading to say that God's word *is* the stone, papyrus, etc. If the stone is destroyed, the word of God is not thereby destroyed, because other copies exist, or if not, God can cause the words to be written again (Exod. 34:1ff.; Jer. 36:27ff.). If we may use the analogy of the Incarnation, the true parallel to Christ's divine and human nature is that God's word is both divine and human language. (By divine language I do not mean an untranslatable angelic language or the like, but language that God speaks.)

The stone is more like the Virgin Mary. It is a bearer of the word, but not identical with the word. The words are not the stone but what is *written* on the stone. What is divine is not the chiseled indentations or the ink, but the literary corpus of words.

Hence I would rather say that God's written word is not a divine-human "being" but divine-human communication. To the subject of communication and other relations I return in 3.3.

2.4 *Creation*

Next, let us look at Creation and see how it can be subdivided. Undoubtedly there are many ways of subdividing, all good. However, it is especially helpful to return to Genesis 1–2. Let us focus on the mandate to Adam in Genesis 1:28-30. Here is man at the very beginning of his task in Creation. He must know something about what he is to do with Creation for the service of God. And God supplies him with the instructions that he needs.

We cannot, of course, know whether God spoke many other things to man besides what is recorded. But Moses undoubtedly wants us to understand Genesis 1:28-30 as a kind of summary, if not the whole, of the instructions that God gave at that time and which continue to hold (though in altered form) even after the fall (see Gen. 3:15-19; 9:1-7).

In the first place, God does not call upon man to have dominion over all Creation, but (so far as I can see) only over those things which are in some sense "within his range." The significant "piece" from 1:1-27 that is missing from 1:28-30 is the sun, moon, and stars (cf. Gen. 9:1-3). The reason is obvious. Man is not expected to have dominion over that which (at least in the beginning) is relatively inaccessible to him. Neither, for that matter, is he expected to have dominion over other men in the way that he has dominion over the animals. To understand what he had to do, man had to understand that there were different parts of Creation to which he had somewhat different relationships. There were (a) heaven, (b) fellow men, and (c) subhuman Creation.

This leads to the following descriptions.

Description. *Heaven* is that part of Creation not accessible to the ruling powers of men made of dust.

Description. The *Cosmos* is that part of Creation which is not Heaven; or, equivalently, which includes men made of dust and subhuman Creation.

Description. *Men* are those Creatures in the image of God called to subdue the earth, have dominion over the animals, etc.

Description. The *Subhuman Kingdom* is that part of nonhuman Creation placed in Genesis 1 under man's dominion and rule.

One can also describe corresponding studies.

Description. *Ouranology, Cosmology, Anthropology,* and *Theriology* are, respectively, the studies of Heaven, the Cosmos, Men, and the Subhuman Kingdom.

Once again, the studies necessarily overlap.

2.41 *Heaven*

The above description of Heaven will probably raise questions. Let me remind readers that in this, as in other descriptions, I do not claim by any means to conform *precisely* to biblical usages. But I have tried to do justice to what I feel is a vagueness in biblical teaching. There is little speculation (least of all in Genesis 1!) about "what exactly is in heaven?" Man does not need to know much about heaven in order to start on his task, and so he is not told. All he needs to know is that the sun, moon, and stars, as well as any other nondivine beings, are creations of God.

The most obvious alternative description would be that "heaven" is that part of Creation above the earth, i.e., above what we walk around on. However, within the horizon of the ancient Near East, this amounts to virtually the same thing as my description. Because of the limitations imposed by "gravity," for the ancients what is *above* the earth is precisely what is inaccessible. This is undoubtedly one reason why it is theologically appropriate for Jesus to take leave of his disciples by ascending (Acts 1:9-11) rather than simply walking away from them or vanishing without motion.

In modern times, of course, traveling upward has become a live option for men. So *now* it becomes relevant to ask wheher one could reach Christ or angels in the same way that one could reach stars. And then it seems to me that the inaccessibility rather than the spatial "aboveness" of the "heaven" of Christ and the angels is what the Bible insists upon.

Obviously I intend that the boundary between Heaven and Cosmos should be vague, as the boundary undoubtedly was for Adam. Birds, for instance, fly "across the expanse of heavens (sky)" (Gen. 1:20; my translation), and are termed "birds of the heavens (air)" (Gen. 1:28, 30; my translation) in the crucial instructions to man. Yet they are placed under man's dominion (1:26, 28), presumably partly because of the fact that they do come down to earth and come "within man's range" in a vague sense. Therefore, they could equally well be classified as Heavenly or Subhuman Creatures, depending on one's viewpoint. I prefer, consistently with my earlier descriptions, to treat them as Subhuman Creatures.[6]

Later on (indeed, soon: Gen. 3:24) man learns about angels. Angels are part of Heaven. For instance, angels are included with "all his host," "sun and moon," "all you shining stars," "you highest heavens," and "you waters above the heavens" in the command to praise the Lord in Psalm 148. This psalm divides neatly into two parts, one headed by "praise the Lord from the heavens" (v. 1), the other by "praise the Lord from the earth" (v. 7). Each heading is followed by exhortations directed to a list of Creatures in the realm in question (vv. 2-4 are about heavenly Creatures, and vv. 7-12 are about earthly [Cosmic] Creatures). Each is terminated by reasons for giving praise (vv. 5-6, 13-14). Hence we may safely conclude that angels are indeed included among the heavenly Creatures (cf. also Matt. 18:10).

Description. *Angels* are personal Creatures who belonged to heaven at the time when they were created.

Description. *Angelology* is the study of Angels.

The other thing to notice about Psalm 148 is that angels and "all his host" (are these personal Creatures?) are included right along with sun and moon, etc., as belonging to heaven. There is no concern to distinguish precisely between the "heaven" of the angels and the "heaven" of the sun and moon. Indeed, the vagueness of biblical language is great enough so that, in our own time, James Reid can try to build a case for the existence of personal Creatures on other planets, on the basis of Nehemiah 9:6; Psalm 148:1-2; Psalm 50:4; Mark 13:27; Psalm 89:5-7, 11; Philippians 2:9-11; Revelation 5:3; Isaiah 24:21.[7] His case I judge highly dubious at best, but it demonstrates how hard it can be to distinguish between the "heaven" and "host" of angels and the visible "heavens" and "host" of stars.

Having imbibed in modern astronomy, we find it hard to place ourselves in the situation of the Old Testament. I strongly doubt whether Israelites thought of "heaven" in a predominantly spatial sense. (Our picture of "space" is no more ancient than Newton.) Angels, stars, and God himself are "in heaven" in the sense of being inaccessible to man's ruling powers. God fills heaven and earth (Jer. 23:23-24), so "heaven" in which God dwells could hardly be simply a "place." At the same time the word 'above' can be used because the whole heaven is described in terms of what is visible of it. This is consistent with the "phenomenal," everyday character of biblical language.[8]

I doubt whether an Israelite ever asked himself whether Psalm 148 was teaching that the angels were really "out there" in the sense that, if you traveled far enough out, you would meet them. That type of question is a question with too much precision. It presupposes a type of speculative interest that the average Israelite did not have. God is not concerned in Psalm 148 to tell us the answer one way or the other. He uses the term 'heaven' as a general, nondistinguishing designation.

A similar kind of imprecision accounts for the lack of any explicit mention of angels in Genesis 1. According to Psalm 148, angels definitely are Creatures (148:5). They are included in a list (148:1-4) whose other members are borrowed from Genesis 1 (especially "you waters above the heavens," an allusion to Genesis 1:7-8). Are the angels, then, to be included in the creation of the heavens in Genesis 1:1? or in 1:7-8? or in 1:14-18 along with the "host" of heaven? or some other place? One cannot be certain.

The only thing that seems certain is that angels were created sometime between the time of Genesis 1:1 and 1:31, when "the heavens and the earth were finished, and *all* the host of them" (2:1). It is necessary also that the fall of Satan and his angels occurred before Genesis 3:1 (cf. Rev. 12:9).

There are at least two possible reasons for lack of angels in Genesis 1.[9] (1) It would have cluttered up an already complex list. Not all the specific kinds of Creatures are listed, but an overview only is given. (2) The emphasis of the chapter is on the creation of those visible Creatures that man sees in everyday life. Microscopic Creatures as well as angels are "left out."

Granted that the term 'heaven' has a range of meaning, is it perhaps still possible to make some distinctions, on a biblical basis, among kinds of "heaven"? What is the significance of "highest heavens" in Psalm 148:4 and the apparent distinction between the "heaven" and "the heaven of heavens" in Deuteronomy 10:14; II Chronicles 2:6; and Nehemiah 9:6? Probably these terms are little more than superlatives meaning "the heavens in all their extent." Or the "highest heavens" may be intended to exclude the lower reaches of the atmosphere. The reader should avoid too quickly making spatial astronomical terminology out of everyday language.

The reader may object to my descriptions that I have thereby forbidden Men to attempt to rule over Heaven. This would exclude not only moon exploration but even investgation of the atmosphere (Gen. 1:20). Actually, however, I have done nothing of the sort. I have not, remember, defined Heaven in primarily spatial terms. The boundaries of Heaven shift, become more precise, or what have you, as Men made of dust gain in technical ability to rule over the earth's atmosphere and the solar system—precisely because Heaven is by definition that part that Men do not yet have access to.

The boundaries are fuzzy, because we can, for example, study the stars to a certain extent, but cannot yet travel to them. The "subduing" in Genesis 1:28 is not a terribly specific term, but in keeping with the context it should most likely be interpreted as focusing on the cultivation and physical transformation of the earth for the service of Men to God's glory. Even in our time, Men are not yet able to do that beyond the moon.

I will not deal with the question of what kinds (if any) of extra-terrestrial science are a wise stewardship of time and resources.

At present, then, most men are limited to the Cosmos. But not all are. Jesus Christ now rules not only earth but also heaven (Matt. 28:18; Eph. 1:21-22). In the future we are to judge angels (I Cor. 6:3). Hence the inaccessibility of Heaven to "men's" ruling powers is confined to men made of dust, who bear "the image of the man of dust" (I Cor. 15:47-49). A temporal and eschatological ("Culmi-national") element enters here, which I will discuss later (3.2).

As a final remark, let me point out that the Bible's teaching about angels is not meant to produce speculation, but (among other things) to remind men that they do not see or know all of the story about Creation. This should put a rein on the dogmatism and overly sweeping conclusions sometimes found in modern science.

2.42 Men

Description. The *Human Kingdom* is all Men taken together.

This kingdom can be subdivided into Men in the Cosmos (who have not yet died) and Men in Heaven. Thus:

Description. The *Cosmic* Human Kingdom is that part of the Human Kingdom in the Cosmos, and the *Heavenly* Human Kingdom is that part of the Human Kingdom in Heaven.

The latter Kingdom is empty until after the fall of man.

It is obvious that an account of Men, perhaps even more than an account about Angels or about the Subhuman Kingdom, will have to include attention to temporal development: creation, fall, redemption, judgment. I will pass over a large part of what the Bible says about Men, because systematic theologies have already covered Anthropology tolerably well. Remarks about Men will be scattered under later topics.

2.43 The Subhuman Kingdom

Now we shall look at biblical subdivisions of the Subhuman Kingdom.

2.431 The three-decker universe

Because I am attempting to construct a biblical Ktismatology, I should at this point say something about the liberal accusation that the Bible contains a "three-decker" mythical cosmological picture. According to this picture, it is said, the universe consists of three layers: "the heaven above, the earth beneath, and the water under the earth" (Exod. 20:4; cf. Phil. 2:10).[10] Related to this is the interpretation that sees Genesis 1 on the background of Near Eastern creation myths.[11]

In response, I limit myself to several cautions and objections.[12]

1. It is sometimes admitted in liberal circles that the supposed three-decker structure is really a pseudo-problem.[13]

2. Biblical language is basically popular language, phenomenal language, describing the world as it appears from man's point of view.[14] For example, language in the Bible about the sun's rising and setting is no more to be interpreted as a scientific theory than is our modern popular language about the sun's motion. Within the universe of discourse of popular languages, to say that the sun rises is true; to say that "the earth turns to make the sun visible" would be false, since it would conjure up some kind of picture in which the observer is stationed on a platform stationary with respect to the sun, while the earth as a blob somehow turns away to make the sun visible.

3. Poetic language is to be used only with utmost caution in drawing cosmological conclusions. Just as a modern poet can talk about the "music of the spheres" without implying approval of Ptolemaic astronomy, so Old Testament poetry can use figurative language, some of which may include allusion to Near Eastern cosmological beliefs.

4. Due allowance must be made for metaphor. "The error is to be avoided of forcing the language of popular, often metaphorical and poetic, description into the hard-and-fast forms of a cosmogony which it is by no means intended by the writers to yield."[15]

5. Biblical cosmology should be based only on what the Bible teaches, not on inferences about what Israelite culture believed. And

this teaching is primarily in sentences and larger units, not in the vocabulary *per se*.[16] (Thus, for example, it is not possible to draw any firm cosmological or cosmogonical conclusion from the supposed etymological relation between the biblical 'deep' (*t^ehom*) and the Babylonia god Tiamat.[17]) I have tried to base my own conclusions not on the bare vocabulary of "heaven" and "earth," but on the different relations which these two have to man.

6. A consistently conservative (i.e., Bible-believing) hermeneutical method must be used. Liberal constructions of biblical cosmology may look plausible from liberals' point of view. However, liberals must realize that rejection of the documentary hypothesis and belief in the genuine antiquity of the creation story carry with them profound alternations away from the typical liberal methods of Old Testament interpretation. For one thing, the account of Genesis 1–2 dates back at least as early as Mosaic times.[18] Hence the poetic commentaries in Proverbs 8:22-31 and elsewhere in Scripture were directed to people who already recognized Genesis 1–2, as Scripture. Proverbs 8 must be seen as deepening some aspects of Genesis 1–2, to be sure, but not as contradicting it, mythologizing it, or offering an alternative.

7. The consistent monotheism and emphasis on the untrammeled divine sovereignty in Genesis 1–2 is an insuperable barrier to interpretation that would see in Genesis 1–2 a real parallel to Near Eastern myths.

8. Superficial parallels between Genesis and other Near Eastern literature can best be accounted for in terms of (a) a possible common source of authentic tradition from Noah or one of Noah's descendants;[19] (b) a polemicizing tendency in Genesis 1–2, which says in effect to the ancient Near East, "Here is the true account of the origin of the world, over against your false polytheistic accounts. God alone is creator."

9. The supposed details in the three-decker view are obvious metaphors. For example, it is said that the heavens have windows through which the rain comes (Gen. 7:11; Isa. 24:18). In another place, however, the Lord has to *make* windows (II Kings 7:2), through which grain comes! In still another case, blessing comes through the

"windows" (Mal. 3:10). Israelites in fact knew that rain comes from clouds (Judges 5:4; Job 36:28f.; Ps. 77:17; 135:7; 147:8, etc.).[20] When the above cautions are observed, the liberal objections virtually disappear. Still, what are we to do with the tripartite division of Exodus 20:4, 11 and Philippians 2:10? It is basically only a refinement of the distinction of heaven and earth (Cosmos) into three parts, heaven, land, and sea (Ps. 69:34; Acts 4:24; Rev. 10:6).[21] But if such is the case, I have no objection to introducing this threefold distinction into modern Ktismatology as well.

On the one hand, the land/sea distinction has not the same crucial significance as the heaven/earth distinction in the laying out of man's task in Genesis 1:28-30. On the other hand, it cannot be denied that the distinction is present: sea, "air," and land animals are explicitly distinguished in Genesis 1:26 and 28, and are listed in the same order as the order of their creation in 1:21, 24-25. At the beginning of his commission man is, as it were, given some sketchy indication of the habitats in which he will meet three major groups of animals.

Description. The *Terrestrial Kingdom* is that part of the Cosmos consisting of the land and its inhabitants.
Description. The *Aquatic Kingdom* is that part of the Cosmos consisting of the waters (seas, rivers, etc.) and their inhabitants.
Description. *Geography* and *Oceanography* are the study of the Terrestrial and Aquatic Kingdoms respectively.

The "fuzzy boundary" between these two Kingdoms includes the amphibians (animal, plant, and mineral).

2.432 *Animal, plant, and mineral*

The question now arises whether Genesis 1:28-30 makes any further distinctions, beyond what we have discussed, that would be of special ontological significance for man. I think that it does. Namely, it distinguishes animal, plant, and inorganic things (see also Gen. 9:1-3). This distinction is drawn in a number of different ways, as Table 1 shows. It was important for man to know this distinction so that he would know something about what was appropriate behavior toward each type of thing—in particular, what he was and

Table 1

The Teaching of Genesis 1:28-30 on Animal, Plant, and Inorganic Kingdoms

sym-bol	manner of distin-guishing	animal	plant	inorganic
P	features	living thing that moves; "living soul" = breather	bearing seed; i.e., growing and reproducing; green	"earth"
W	the task of man with respect to the Creature in question	have dominion	have them for food	fill and subdue
F	relations to other King-doms	eating plants; moving on the earth	food for animals; growing on the earth	support for animals and plants

was not to eat.[22] Hence this distinction was a rather basic distinction for Adam's world. Furthermore, Genesis makes an obvious distinction among the three kingdoms in 1:11-13, 14-19, 20-25.

Therefore, we obtain the following descriptions.

Description. The *Animal Kingdom,* the *Plant Kingdom,* and the *Inorganic Kingdom* are the divisions of the Subhuman Kingdom laid out for man in Genesis 1:28-30. They consist respectively of animals (characterized as moving and breathing), plants (characterized as green, growing, and reproducing), and nonliving things (earth). The studies of these Kingdoms are respectively *Zoology, Botany,* and *Inorganics.*

Description. A *Kingdom* is any one of these three Kingdoms or the Human Kingdom.

Obviously, there is a vagueness about the exact borders of these Kingdoms. For convenience, we might want to describe the plant/

Table 2

Summary of Ontology

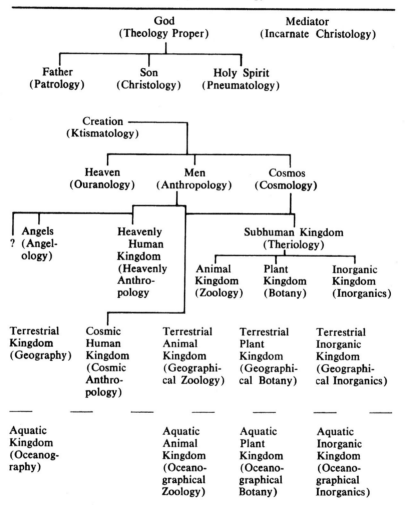

(Names of studies are listed in parentheses directly under the subject-matter of study.)

nonliving-thing distinction in terms simply of living/non-living—but this would be little better than the above description. To what Kingdom do viruses belong? Again, the animal/plant distinction could be drawn in terms of lack of chlorophyll. But this creates problems with fungi and plant spores, and it is open to the larger objection that it describes animals in terms of a *lack* rather than in terms of the *positive* abilities that are the most distinctive mark of common animals. I am not saying that any one possible description is *the* right one, but simply pointing out that vagueness of one sort or another is unavoidable.

Of course, it would be perfectly possible to introduce into our discussion more of the distinctions from Genesis 1:28-30 (e.g., herb/tree?), but there would be no particular purpose in doing so. Certainly Genesis makes no claim that these are the only real distinctions that we can make, or that any distinctions cutting across these are somehow "invalid," or that these distinctions will always prove to be the most useful ones in any human endeavor. There are many ways of "cutting the cake," as the variation in description even in the course of Genesis 1:28-30 shows. Yet, taking all these provisos into account, the distinctions that are described above (apart, perhaps, from Terrestrial/Aquatic) have crucial significance for setting man on his way to fulfilling the task that God has given him. We may summarize the ontology at which we have arrived in Table 2.

NOTES TO CHAPTER 2

1. See further John M. Frame, "The Word of God and the AACS; a Reply to Professor Zylstra," *Presbyterian Guardian* 42 (Apr., 1973), p. 60.

2. See John M. Frame, "What Is God's Word," *Presbyterian Guardian* 42, (Nov., 1973), p. 142.

3. H. Evan Runner, *Syllabus for Philosophy 220: The History of Ancient Philosophy* (unpublished; Grand Rapids: Calvin College, 1958–1959), p. 18.

4. Abraham Kuyper, *Encyclopedia of Sacred Theology* (London: Hodder and Stoughton, 1899), pp. 228ff.

5. Christ was mediator in creation and in OT revelation and redemption. But he was not yet *Mediator* in my technical sense of incarnate. See further 3.323.

6. According to my argument in the paragraph following this one, Ps. 148 classifies birds with the Cosmos rather than with Heaven (148:10). This is

a confirmation of the appropriateness of my classification of birds. Cf. Gen. 1:22, and Edward J. Young's statement: "The sphere in which the birds are to live is said to be the earth, not the firmament." *Studies in Genesis One* (Philadelphia: Presbyterian and Reformed, 1964), p. 72.

7. Reid, *God, the Atom, and the Universe* (Grand Rapids: Zondervan, 1968), pp. 210ff.

8. See 2.431.

9. I regard as conclusive the arguments by Bernard Ramm against the gap theory (which postulates that Gen. 1:3ff. is a re-creation from a situation of judgment on fallen angels in 1:2). See Bernard Ramm, *The Christian View of Science and Scripture* (Grand Rapids: Eerdmans, 1954), pp. 201-210.

10. E.g., Rudolf Bultmann, "New Testament and Mythology," *Kerygma and Myth*, ed. Hans Werner Bartsch (London: S.P.C.K., 1957), p. 1. The liberal view goes back eventually to Giovanni Schiaparelli, *Astronomy in the Old Testament* (Oxford: Clarendon, 1905).

11. T. H. Gaster, "Cosmogony," *The Interpreter's Dictionary of the Bible*, ed. George Arthur Buttrick (New York: Abingdon, 1962), I, 702-706; Samuel R. Driver, *The Book of Genesis*, 11th ed. (London: Methuen, 1920).

12. For a fuller discussion, see Ramm, *Science and Scripture*, pp. 96-102.

13. Austin Farrer, *Kerygma and Myth*, ed. Hans Werner Bartsch, p. 216. Liberals have also realized that the predominant emphasis of the OT is on cosmo*gony* (origins) rather than cosmo*logy* (structure). See, for example, Gaster, "Cosmogony," p. 702b.

14. "The view taken of the world by Bib. writers is not that of modern science, but deals with the world simply as we know it—as it lies spread out to ordinary view—and things are described in popular language as they appear to sense, not as telescope, microscope, and other appliances of modern knowledge reveal their nature, laws and relation to us." James Orr, "World," *International Bible Encyclopedia* (Grand Rapids: Eerdmans, 1939), V, 3106a. Orr also quotes Calvin: "Moses wrote in the popular style, which, without instruction, all ordinary persons endowed with common sense are able to understand. . . . He does not call us up to heaven; but only proposes things that lie open before our eyes"—*ibid.*, 3108a, cited from John Calvin, *Commentaries on the First Book of Moses Called Genesis* (Grand Rapids: Eerdmans, 1948), pp. 86, 87. Cf. Ramm, *Science and Scripture*, pp. 65ff.

15. Orr, *ISBE*, V, p. 3106a.

16. James Barr, *The Semantics of Biblical Language* (London: Oxford 1961), pp. 234, 263, 249, 269.

17. For a discussion of this and other aspects of the Gen. 1-2 account, see Walter C. Kaiser, Jr., "The Literary Form of Genesis 1-11," *New Perspectives on the Old Testament*, ed. J. Barton Payne (Waco, Tex.: Word, 1970), pp. 48-65; Edward J. Young, *Studies in Genesis One* (Philadelphia: Presbyterian and Reformed, 1964), pp. 15-42.

18. Of course, I do not exclude the use of earlier sources by Moses. See J. S. Wright and J. A. Thompson, "Genesis, Book of," *The New Bible Dictionary*, ed. J. D. Douglas (London: Inter-Varsity, 1962), pp. 460b-461a.

19. "One can also judge that the events of Gn. 1-11 are historical. That

being the case, it would not then be at all surprising if the story concerning them should come to be mythologized in pagan traditions, while being preserved in authentically historical form within the stream of tradition of which Gn. 1–11 is the inspired deposit" —Meredith G. Kline, "Genesis," *The New Bible Commentary Revised,* ed. D. Guthrie *et al.* (Grand Rapids: Eerdmans, 1970), p. 79b. Cf. also discussion of the possibility of genealogical-history sources in R. K. Harrison, *Introduction to the Old Testament* (Grand Rapids: Eerdmans, 1969), pp. 543-553.

20. Orr, *ISBE*, V, p. 3106b.

21. Samuel Jackson, ed., *The New Schaff-Herzog Encyclopedia of Religious Knowledge* (Grand Rapids: Baker, 1953), XII, 428a.

22. Does Gen. 1:29-30 imply a (temporary) *prohibition* of the use of animal food by man? We cannot tell. Perhaps for reasons unknown to us, Gen. 1 simply omits mention of such food. Whether the eating of animals began before the fall or in connection with Gen. 3:21; 4:2; 9:3; or some unrecorded incident, seems impossible to determine with certainty.

Chapter 3

METHODOLOGY

Now I am ready to deal with the second question basic to philosophy of science, namely the question of methodology. But perhaps 'methodology' is a poor word for it. It makes one think of the techniques (a) that scientists actually use and (b) that scientists "ought" to use in the judgment of someone. Now, I do not want to exclude completely either one of these questions, but for the moment I want to ask a much broader question: how does everything function? This question is not asking only about scientists, or only about men, but also about the Subhuman Kingdom.

The question "how does everything function?," like the question "what is there?" considered earlier, is a somewhat vague, puzzling, and intractable question. Again one asks, "Just what kind of answer would satisfy the inquirer?" Again, one could try to answer facetiously by explaining more or less at random how to cook various foods, how to repair machinery, what to do when you get sick, and so forth. Or, one can treat the question as a religious question that is searching for some *one* comprehensive answer to why the world is the way it is. A suitable answer is "Jesus Christ bears all things along by his word of power" (Heb. 1:3; my translation).

If, now, we want to become more specific, there are various ways of proceeding, any one of which might be faithful to the Bible's teaching. (The Book of Proverbs is one possible approach.) However there is a danger of leaving out or distorting elements of the Bible's teaching, if we are not careful to distance ourselves from the sins of our heart and the corruptions of secular thinking in our time. There is no simple method of guarding against this danger, apart from the purification of our whole selves by the Lord.

27

Having said all this, however, I believe that the simplest procedure is a return to a further examination of Genesis 1:28-30. I have already argued that this passage has a kind of foundational significance for giving man his preliminary orientation in Creation. By returning to it I obtain the additional advantage of being able to integrate what I now discuss with the previous discussion of ontology. Of course, I could not avoid saying some things about "methodology" in chapter 2. Neither can I here avoid saying things about "ontology" at the same time that I talk about "methodology." After all, we must talk about God and his Creation or not talk at all. Thus the subjects of discussion of this book all interlock with one another.

Because of the differences among man, animal, plant, and inorganic Creature, a discussion of functions will, to a certain extent, have to deal with the four Kingdoms (Human, Animal, Plant, Inorganic) separately. In accordance with Table 1, I divide the discussion into three sections: (3.1) a discussion of the more or less constant characteristics of the Kingdoms (P of Table 1: "modality"); (3.2) a discussion of the historical development of the Creation under the direction of Men (W of Table 1: "temporality") (3.3) a discussion of connections and relations of Creation (F of Table 1: "structurality"). Naturally, these three sections will interlock with one another, rather than being water-tight compartments.

3.1 *Modality*

By a discussion of the characteristics of Kingdoms, I hope to obtain the beginnings of a classification of sciences. Different sciences are, at least to a degree, interested in different characteristics of Creatures. Hence an analysis of the different characteristics of the Kingdoms will help us to explain and appreciate the diversity among sciences.

In the subsequent discussion, the broadest classificatory areas will be called "Modes" (3.11) and less broad areas "Functions" (3.12).

3.11 *Modes*

According to Genesis 1:28-30, the characteristics of the Animal, Plant, and Inorganic Kingdoms are as follows. (1) Animals live, move ("creep"), breathe. (2) Plants bear seed, grow and reproduce, and are green. (3) The "earth," by implication, is the nongrowing, non-living platform for animals and plants. To this list we could add the characteristics of man: he rules, subdues, is fruitful, reproduces ("multiply").

Of course, each one of these lists is a preliminary sketch rather than an exhaustive categorization. Hence we may proceed, if we wish, to add further characteristics helpful for identifying the uniqueness of each Kingdom. However, we must be careful. Note that several characteristics listed in Genesis 1:28-30 are, in some sense, *not* unique to the Kingdom in question. Men and animals as well as plants grow and reproduce. Men as well as animals move about and breathe (Gen. 2:7 KJV "living soul" is a translation of the same Hebrew phrase as occurs in 1:30, probably meaning "a breather"; cf. Gen. 9:10, 12).

If we may generalize, men do some things that animals do not (speaking, ruling,[1] buying and selling, worshiping), animals do some things that plants do not (moving about, breathing, fearing [Gen. 9:2]), and plants do some things that Inorganic Creatures do not (growing, reproducing). There is a kind of order from higher to lower: man, animal, plant, Inorganic Creature. This order is manifested (a) in the simple description of the relation between Kingdoms (man has dominion over animal, animal eats plant, plant grows on the earth), where the lower serves the higher. And the order is further manifested (b) in the new capabilities of the higher in comparison with the lower. On the other hand, the major characteristics common to the lower are *also* to be found in the higher. For example, not only the Inorganic Kingdom but all the Cosmos has color, shape, texture, temperature, and so on.

Description. A *Mode* is the bundle of characteristics that a Kingdom has in addition to those of lower Kingdoms. The *Personal* Mode, the *Behavioral* Mode, the *Biotic* Mode, and the *Physical* Mode are the names of the Modes of the Human, Animal, Plant, and Inorganic Kingdoms respectively.

Description. *Ethology, Behaviorology* (or *Praxeology*), *Biology,* and *Physics* are, respectively, the studies of these four Modes.

See Table 3.

Table 3

A. Modes

Kingdom	Mode	Study	Examples
Human	Personal	Ethology	ruling, speaking, buying
Animal	Behavioral	Behaviorology	breathing, fearing, eating
Plant	Biotic	Biology	living, growing, reproducing
Inorganic	Physical	Physics	having a color, shape, temperature, weight

B. Overlapping Studies

	Inorganics	Botany	Zoology	Anthropology	Ouranology	Theology Proper
Ethology studies the Personal Mode of				Men	Angels; Heaven; Heavenly Men	God
Behaviorology studies the Behavioral Mode of			Animals	Men	Heaven	God
Biology studies the Biotic Mode of		Plants	Animals	Men	Heaven	God
Physics studies the Physical Mode of	Inorganic Creatures	Plants	Animals	Men	Heaven	God

It seemed to me less dangerous to coin terms ('ethology,' 'behaviorology') than to use English terms already freighted with too much connotation that I do not want. I remind readers again that *none* of my technical terms have precise boundaries.[2]

Ethology, Behaviorology, Biology, and Physics cut across the earlier divisions into Theology Proper, Ouranology, Anthropology, Zoology, Botany, and Inorganics (see Table 3B). Biology, for example, studies the Biotic Mode of Plants, and thus overlaps with Botany. It also studies the Biotic Mode of Animals (their living, growing, etc.), and thus overlaps with Zoology. And so on.

But not all the studies overlap so directly. For example, since Plants do not have the Personal characteristics of ruling, speaking, and so on, Ethology does not directly overlap with Botany (note the blank space in Table 3B). This does not imply, of course, that we will not ever talk about Plants in discussing Man's responsibilities in ruling, speaking, and so on. Hence Ethology and Botany *do* actually interlock with one another, though in a more indirect fashion than, say, Biology and Anthropology (which have a common interest in the living, growing, and reproducing of Men).

Can Ethology, Behaviorology, Biology, and Physics include the study of Heaven and God? It seems clear that Heaven and God do have *some* of the characteristics of Cosmic Creatures. For example, angels speak, rule, and wage war (Dan. 10:13), and thus are persons (but do they buy and sell?); they "come and go" in a way that shows activity in the Behavioral Mode (Dan. 10:13-14); they are living (on the other hand, they do not marry or reproduce—Luke 20:36); when they appear to men, they have color, shape, number, and so on (but one wonders how far they retain these characteristics apart from their appearing to men—Luke 24:39). The stars, at any rate (if one wishes to classify them as Heavenly), have a Physical Mode.

About God similar statements can be made. God speaks, rules (Personal Mode), is angry or pleased (Behavioral Mode), lives (Biotic Mode), and fills all things (Physical Mode). However, one must be cautious about pressing the similarities. God does not marry (Personal), breathe air (Behavioral), reproduce (Biotic), or have color or shape (Physical). And one could go on to say that God does not speak in the same *way* that men do (he does not have a body, and his speaking has *divine* authority); God is not angry in the same *way* that men and animals are (he is not changeable, and his anger is always a perfect, righteous anger against sin); and so on. Table 3B is not intended to deny these factors.

We must now ask, should two more "Modes" be added, one for Angels and one for God? These two "Modes" would then presumably include (1) those characteristics that Angels have in addition to Personal characteristics, and (2) those characteristics that God shares with no Creature. I have no objection in principle to adding these two "Modes." However, there is a difficulty, in that God and angels are inaccessible to men's powers of rule. In the case of angels, the result is that it is difficult to say anything about what those additional "characteristics" might be. Angels can perhaps do some things *better* than men, but it is difficult to come up with anything convincingly remote from what man might be able to do. Hence, as far as I am concerned, the supposed "Angelic Mode" is a complete blank at least until we die. To avoid speculation, I have left it out.

In the case of God, we *can* say something about what God can do and man cannot. For example, God is not subject to change in time, the Father begets the Son, and God created the world. So one might want to speak of a "Divine Mode." However, I will not, because it seems to me as much confusion-producing as helpful. In the first place, there is a great deal of mystery in any human speech about what is absolutely unique to God. Even in the above examples I have had to use words that are often applied to men: "change," "beget," "create" (e.g., an artist creates a picture).

In the second place, everything that God does is, in a sense, *unique* to him. When he speaks, his speech is divine speech, with divine wisdom, holiness, and power; his life is divine, self-sufficient, underivative life; likewise his immensity (filling all things) is such that *God* (not a "part" of him) is present everywhere. Hence talking about a "Divine Mode" is not really different than simply talking about God. The study of the "Divine Mode" is Theology Proper.

In other words, I am saying that (a) as far as our human vocabulary is concerned, the four Modes pretty well cover the ways we have of speaking about things; that (b) some peculiarities occur in Modal study of God, because of the Creator/Creation distinction.

3.12 Functions

Now I wish to look at further distinctions within Modes.

3.121 Functions within the Personal Mode

Because a great variety of activities are included under the Personal Mode, I will attempt some subclassification. This will, moreover, prove useful in understanding *Man* the scientist who is always the actor behind scientific results.

3.1211 F: ordinantial Functions

First, the Cosmic Human Kingdom cannot be broken up into parts in the way that the Animal Kingdom was separated from the Plant Kingdom. Some other method of classification will have to be used. And Genesis 1:28-30 is too brief a passage to form a basis for further classification. So let us proceed further into the Bible. What kind of responsibilities are given to man in more detail in Genesis 2? He is to hallow the sabbath (2:3),[3] to till and keep the garden (2:15), and to care for his spouse in marriage (2:23-25). In addition, of course, God gives the special "probationary" command with respect to the tree of knowledge (2:16-17). Finally, there are hints of more specialized areas of endeavor, like economics (2:11-12), taxonomy (2:19-20), and aesthetics (2:9).

Leaving the probationary command apart, man's calling falls in three areas: sabbath, family (including marriage and parent-child relations), and labor. These are the three "creation ordinances."[4] They involve man's relation to God (sabbath), to the Cosmic Human Kingdom (family), and to the Subhuman Kingdom (labor). Nevertheless, on closer examination all three ordinances involve all three relations. For example, man's care for his family and his labor must be to the glory of *God*. Man's labor provides food, greater understanding (naming the animals), and beauty (keeping the garden) for himself and others. His sabbath rest is for *his* own good (Deut. 5:14; Mark 2:27). And so forth. The three ordinances interlock with one another.

Description. The *Sabbatical Function* consists of that part of the Personal Mode having to do with activities normally characteristic of the sabbath ordinance, or equivalently, of a person's direct relation to God. The *Social Function* consists of that part of the Personal Mode having to do with activities normally characteristic of the ordinance of family, or equivalently, of a person's relation to the Cosmic Human Kingdom. The *Laboratorial Function* consists of that part of the Personal Mode having to do with activities normally characteristic of the ordinance of labor, or equivalently, of a person's relation to the Subhuman Kingdom.

Description. *Liturgiology, Sociology,* and *Ergology* are, respectively, the studies of the Sabbatical, Social, and Laboratorial Functions.

This does *not* say, e.g., that Sociology is the study of Man's relation to Man, or of the ordinance of the family. There would be nothing wrong with such a definition, but it is not what 'Sociology' means. "Sociology" studies those characteristics of Man that come to the fore in family relations, but it will study those characteristics as they occur also in God, and in Man's relation to the Subhuman Kingdom. Thus we call a Man's *speaking* to his dog Social, even though he is speaking to his *dog*. Elsewhere (2.4) the dissimilarity between a dog and a man has been emphasized; here the emphasis is on the similarity between the two kinds of speech. Again the boundaries are fuzzy. Are a dog's yelps "Social"?

Furthermore, the above descriptions do not mean that Sociology studies the Social Function somehow "in the abstract"; rather, it studies God and Creation, focusing on those types of activity included in the Social Function (see Table 4). The large overlap among the Functions is evident from our vocabulary. For example, ordinary English (or Hebrew or Greek) does not possess all *that* many specifically Sabbatical terms. Instead the relations of God to man are expressed using Social and Laboratorial terms. As a result, Sociology will include study of the Social Function of God., i.e., of God's loving, commanding, telling, asking, ruling, punishing, providing, etc. Ergology will include study of the Laboratorial Function of God, i.e., of God's thinking, creating, appreciating, etc. Liturgiology, on the other hand, studies not only Man's but also God's prophesying,

Table 4

Ordinantial Functions

Function	Basic Idea	Study	Examples
Sabbatical	to God	Liturgiology	pray, worship, preach, evangelize, bless, curse
Social	to Men	Sociology	command, greet, tell, ask, speak, serve, obey, punish inherit, buy, welcome
Laboratorial	to the Subhuman Kingdom	Ergology	remember, think, build, weigh, expect, appreciate

blessing, cursing, etc. The difference is that in Liturgiology everything is viewed in the light of the *uniqueness* of the sabbatical ordinance rather than in terms of the similarity of God-Man relations to Man-Man relations or Man-Creature relations.

3.1212 P: official Functions

We can analyze man's task not only in terms of the three creation ordinances, but also in terms of three offices: prophet, king, and priest. These terms, of course, originate primarily in the Mosaic period. But the functions of prophet, king, and priest are by no means confined to the period. Jesus Christ is our final (eschatological) prophet, king, and priest. The Book of Hebrews, especially, asserts this threefold office at its very beginning: "God spoke of old . . . by the *prophets;* but in these last days he has *spoken* to us by a Son, . . . *upholding* the universe by his word of *power.* When he had made *purification* of sins, he *sat down at the right hand* of the Majesty of high, . . ." (1:1-4); the rest of Hebrews confirms this introduction.

Now, Jesus Christ as federal head of his people is the last Adam (I Cor. 15:45). Hence we expect that he is appointed to offices in order to restore the offices that Adam perverted by the fall. Sure enough, one can detect a "seed" form of the offices of prophet, king, and priest in Adam. Adam listened to the word of God (Gen. 1:28-

30) and named the animals (2:19-20, 23; prophetic); he had dominion over the animals (1:26; kingly or basilic); and he had
communion with God in the garden (3:8), and received the promise
of final blessing or cursing (2:9, 16-17; priestly or hieratic).

Thus it is possible to trace prophetic, kingly, and priestly functions
all the way through the Bible. As with the Sabbatical, Social, and
Laboratorial Functions, there is an enormous amount of diversity and
interlocking among these offices. Except in Genesis 1–2, of course,
we see these offices as they have been shaped and transformed by the
fall and the history of redemption after the fall. Some things, like
the offering of bloody sacrifices by Old Testament priests, are characteristic not of all priests (cf. I Pet. 2:5; Heb. 13:15), but of the
Levitical priesthood and the final offering of Jesus Christ (Heb.
9:14; 12:24). Presumably there would have been no need for bloody
sacrifices apart from the fall.

What, then, is the element of continuity in these three offices?
There need not be only *one* such element. However, if I were to
sum up what was distinct to each office, in contrast to the rest, I
would say that (a) the prophet has to do with meaning, communication, wisdom, information; the king with rule, power, mastery; the
priest with sharing and communion, especially communion of goods,
of blessing or cursing (see, e.g., Num. 6:22-27, and the picture of
communion in blessing throughout the Book of Hebrews [e.g., Heb.
7:25; 9:14, 15]).

> Description. The *Prophetic, Kingly,* and *Priestly* Functions are
> those parts of the Personal Mode having to do with activities,
> states, characteristics, etc., of a predominantly prophetic, kingly,
> and priestly sort, respectively.

Or, one can say,

> Description. The *Prophetic* Function has to do with meaning,
> communication, wisdom, information; the *Kingly* or *Basilic* Func-
> tion has to do with rule, power, mastery; the *Priestly* or *Hieratic*
> Function has to do with communion.

> Description. The studies of the Prophetic, Kingly, and Priestly
> Functions are *Prophetics, Basilics,* and *Hieratics.*

See Table 5.

Table 5

Official Functions

Function	basic idea	Study	examples
Prophetic	meaning, communication, wisdom, information	Prophetics	pray, preach, evangelize, command, greet, tell, ask, speak, remember, think
Kingly, Basilic	rule, power, mastery	Basilics	worship, serve, obey, build, weigh
Priestly, Hieratic	communion, especially of goods, blessing (or cursing)	Hieratics	punish, inherit, buy, welcome, expect, appreciate, bless, curse

The three Functions overlap and interlock, first of all in the sense that some biblical offices have a kind of combination of two or more Functions. For example, the judge in Deuteronomy 17:12 exercises Kingly functions in conjunction with the priests; Ezra the priest exercises a Prophetic function in Nehemiah 8:5. Samuel seems to be a prophet, king, and priest (offering sacrifice) all rolled into one. The Functions also interlock in the sense that all three offices involve one another at every point. When a prophet speaks the word of the Lord (Prophetic), he also proclaims, asserts, and advances the sovereign rule of God (Kingly), and the hearing of the word brings either the blessing of communion or the cursing of separation (II Cor. 2: 15-17). Thus the Functions deal with matters of emphasis; there should be no attempt to rigidly separate them.

I should add (if it is not obvious by now) that I use 'Prophetic,' 'Kingly,' and 'Priestly' in a very broad sense, including thereby not only the special and extraordinary offices of the Bible, but the everyday speaking, ruling, and fellowship of Men. Thus, for example, a man is a Prophet not only when he speaks the word of God under inspiration (II Pet. 1:21), or in an official capacity (Eph. 4:11; II Tim. 4:2), but whenever his words are seasoned with the salt of helpfulness (Col. 4:6; Eph. 4:29, 31; 5:4; James 3:5ff.; Prov. 10: 11, 13, 14, 19-21, etc.). Even foolish speech is but a perversion of the original Prophetic calling. It is a kind of "false" Prophecy.

I Peter 2:9 comes near to this kind of broadness: "You are a chosen race, a royal [Kingly] priesthood [Priestly], a holy [Priestly] nation [Kingly], God's own people, that you may declare [Prophetic] the wonderful deeds of him who called you. . . ."

If, now, the Prophetic (Pr), Kingly (Ki), and Priestly (Hi) Functions are "intersected" with the Sabbatical (Sa), Social (So), and Laboratorial (La) Functions, we obtain nine more "specialized" Functions, each of which can be described as that part of the Personal Mode covered by two Functions.

> Description. The *Dogmatical* (Pr,Sa), *Presbyterial* (Ki,Sa), *Diaconal* (Hi,Sa), *Lingual* (Pr,So), *Juridical* (Ki,So), *Economic* (Hi,SO), *Cognitional* (Pr,La), *Technical* (Ki,La), *Aesthetic* (Hi,La) Functions are the nine Functions so obtained.

> Description. *Dogmatics, Presbyteriology, Diaconology, Linguistics, Jurisprudence, Economics, Logic, Technology,* and *Aesthetics* are the studies of these Functions, respectively.

See Table 6.

I must warn readers not to place too much stock in the words chosen for these Functions. 'Economic,' 'Aesthetic,' and 'Logic' are probably the least satisfactory terms, because I am using them in a much broader sense than is customary in English. It helps to keep in mind the etymological significance of these three terms, or to think of "Economics" in terms of sharing and providing (not merely material things), "Aesthetics" in terms of appreciating and appraising, "Logic" in terms of studying and cognition.

Note that Technology is the *study* of, not the *practice* of, making, fixing, building, exploiting, mastering, etc.

3.1213 *W: actional Functions*

I am inclined in some ways to stop my analysis of Functions at

Table 6

Personal Functions

official Functions ordinantial Functions	Prophetic Prophetics	Kingly Basilics	Priestly Hieratics
Sabbatical Liturgiology	Dogmatical Dogmatics pray, preach, evangelize	Presbyterial Presbyteriology worship, excommunicate	Diaconal Diaconology bless, curse
Social Sociology	Lingual Linguistics command, greet, tell, ask, speak	Juridical Jurisprudence serve, obey, rule	Economic Economics punish, inherit, buy, welcome
Laboratorial Ergology	Cognitional Logic remember, think	Technical Technology build, weigh	Aesthetic Aesthetics expect, appreciate

(Functions are listed, followed by the name of the study of the Function, followed by examples within the Function.)

this point. Up till now I feel that I have been able to justify my procedure in a fairly convincing way from Scripture. Beyond this point, my further steps in describing supposed "Functions" will be more speculative. I feel that for that reason, if no other, they are less important. Certainly my total exposition does not stand or fall according to the validity of my more speculative steps. However, I do not feel embarrassed about having to speculate. The Bible does not claim to tell us everything that we might want to know about the world, but enough to give us a God-honoring orientation both to what we know and to what we do not know.

Besides this, I do not claim that the further descriptions that I will give are the only way or even the best way of making further distinctions within Modes. They are *one* way, and I think a useful way. What is "best" depends on what one is trying to accomplish.

The distinction that I am thinking of is one among Active, Middle, and Passive Functions. But I am not using the words Active, Middle, and Passive in a grammatical sense. I want these three terms to talk about different (but overlapping) kinds of activity. For example, under the Juridical Function, one can rule (Active), cooperate (Middle), or obey (Passive). Under the Lingual Function, one can speak (Active), discuss (Middle), or hear (Passive).

It is easy to find traces of this distinction in Scripture. For example, the difference between the extraordinary or special office of king (e.g., David or Solomon) and the ordinary office exercised by all members of the covenant community is largely a matter of the more Active character of the Kingly Function in the case of the extraordinary office. The terminology in the systematic theology of the "active" and "passive" obedience of Christ is closely related to (though not identical with) the distinction that I am making. However, I do not insist too much on such apparent biblical roots.

> Description. The *Active* Function is that part of the Personal Mode having to do with activities and characteristics where the persons in question take some kind of initiating role, where they are giving, where they are affected, as it were, from inside out. The *Passive* Function is that part of the Personal Mode having to do with activities and characteristics where the persons in question take some kind of responding role, where they are receiving, where they are affected, as it were, from outside in. The *Middle* Function is that part of the Personal Mode having to do with a mutual interchange, a sharing.
> Description. *Energeticology, Mesology,* and *Patheticology* are respectively, the studies of the Active, Middle, and Passive Functions.

Frankly, I do not think I have done a very good job in describing these three Functions. Readers can probably obtain a better idea of what I mean by examining which entries in Table 7 are classified under each of the three Functions.

3.122 *Functions within the non-Personal Modes*

To show some of the potential of the distinctions already developed, I now explore some possible subdivisions in the non-Personal Modes.

This and the succeeding section (3.123) continue to be somewhat speculative, and hence less important.

It seems possible to me to detect in the Behavioral, Biotic, and Physical Modes some foreshadowings or adumbrations of the Prophetic, Kingly, Priestly, Active, Middle, and Passive Functions. One should really not be surprised at this possibility. Higher animals, especially, engage in many activities very similar to human activities. In this case it should be fairly easy to identify some kinds of activity that remind one of the Prophetic Function, the Kingly Function, and so forth.

For example, in some activities animals are more active (running, jumping, looking, watching, finding, etc.), and in others they are more passive (sitting, lying, sleeping, seeing, hearing, etc.). Still other activities have an intermediate status (perceiving, observing, enjoying, etc.). Hence we may divide animal characteristics into three parts, active, passive, and middle. But the Active, Passive, and Middle Functions have already been described as subdivisions of the *Personal* Mode rather than of animal characteristics. To avoid confusion, we need a new set of terms for the non-Personal Modes. I introduce the term 'adumbrative' (shadowy) to cover the new cases.

Description. The *Adumbrative Active Function* of the Behavioral Mode consists of those activities and characteristics of the Behavioral Mode where, as a rule, the actor takes greater initiative. Similarly, the *Adumbrative Passive Function* and *Adumbrative Middle Function* consist of those activities and characteristics of, respectively, a passive and an intermediate, reciprocal sort.

Similar subdivisions (somewhat harder to detect) may possibly be found in the Biotic and Physical Modes. In that case, one may speak of the Adumbrative Active, Passive, and Middle Functions of these Modes.

What about the Prophetic, Kingly, and Priestly Functions? Do these have adumbrative analogues? Perhaps. I have supposed that the Prophetic, Kingly, and Priestly Functions have to do most intimately with meaning, power (execution), and value. In an animal, one detects this trio in its perceiving (seeing, smelling, etc.: meaning), its acting (running, jumping, crawling: power), and feeling (being angry, afraid, comfortable, torpid: value). In plants, a certain kind of "communication" may be observed in plants' response to changing environment and in the interplay of different plant parts. The "dynamic" or "power" of a plant is perhaps best manifested in growth. And its "value" is manifested in the difference between sickness and health, in the process of repair, reproduction, and death (see Table 7).

Description. The *Adumbrative Prophetic, Kingly*, and *Priestly Functions* are those parts of the non-Personal Modes having to do with meaning, power, and value respectively.

In general, we may summarize as follows.

Description. The *Adumbrative Prophetic, Kingly, Priestly, Active, Middle,* and *Passive Functions* are the adumbrative forms of these Functions found within the non-Personal Modes.

As a special case of this, we have the following.

Description. *Mathematics, Kinematics,* and *Energetics* are, respectively, the studies of the Adumbrative Prophetic, Kingly, and Priestly Functions of the Physical Mode.

3.123 *Views*

Up to now, in discussing the subdivisions of what there is, I have left the role of the human observer, analyzer, and evaluater comparatively in the background. His role has been implicit rather than explicit. But for certain purposes it is useful to introduce *explicitly* the role of human perspectives or "views," in order that we may "cut the cake" in still other directions.

First, Men have the power to focus their attention on any of an indefinite number of items.

Description. An *Item* is anything that Man selects for notice or study.

Thus the word 'Item' will be the broadest technical term that we have (but hardly the most important). It includes not only God and Creatures, but words, Functions, institutions, events, activities, relations, etc. These things are Items when they become the focus of human attention.

Men can study an Item from any of an indefinite number of perspectives or views. For example, suppose that a house is chosen as the Item for study. First, the house may be viewed as a "chunk" distinct from its environment, identifiable by certain features. It may be identified as being a house rather than a shed, a garage, or a store. It may be identified by street address, color, shape, number of rooms, and so on. Let us call this a "particle" view of the house.

Second, the house may be viewed in continuity with its surroundings, as flowing into its environment rather than as a "chunk." It is in the process of repair or decay. Paint chips off or is painted on. Dirt is blown in or swept out. The house becomes empty or full of inhabitants and furnishings. Let us call this a "wave" view (dynamic view) of the house.

Third, the house may be viewed as one element in a system of interlocking relations. The house is one element in a spatial array of houses on the block. It is an element in an array of house designs drawn up by architects according to cost, convenience, and size criteria. It is one element in a temporal series of environments that a man may enter during one twenty-four hour period; and each element in the series is correlated to functions and activities that a man may best perform in the environment. He sleeps and eats in the house, but (typically) earns money elsewhere. And so on. Let us call this a "field" view (relational view) of the house.

These three kinds of view are summarized in the following description.

> Description. A *Particle View* by a Man is a focusing on Items with their closure properties, including an ordering of Items in a taxonomy according to some set of features convenient for the purpose in hand. A *Wave View* is a focusing on sequence characteristics of Items, not requiring sharp segmentation at the borders of Items. A *Field View* is a focusing on the interdependent characteristics and relations of Items.[5]

This book contains some further examples of the use of these three Views. 3.1211, 3.1212, and 3.1213 are, approximately speaking, Field, Particle, and Wave Views of Functions. On a larger scale, 3.1, 3.2, and 3.3 are Particle, Wave, and Field Views of Methodology.

Table 7

Display of Functions

Personal:	Prophetic	Kingly	Priestly
	Ac give meaning	Ac empower	Ac commune
	Pa receive meaning	Pa be empowered	Pa be communed with
Sabbatical	Dogmatical	Presbyterial	Diaconal
	Ac dogmatize, pray	Ac worship	Ac minister, offer offerings
	Pa hearken	Pa submit (religiously)	Pa receive divine blessing
Social	Lingual	Juridical	Economic
	Ac speak	Ac rule	Ac provide
	Mi discuss	Mi cooperate	Mi share, bargain
	Pa hear	Pa obey, serve	Pa receive
Laboratorial	Cognitional	Technical	Aesthetic
	Ac characterize, name	Ac make, fix, build	Ac decorate, adorn, sanction
	Mi study	Mi use, exploit	Mi appraise
	Pa know, understand	Pa master	Pa appreciate
Behavioral:	PrBehavioral	KiBehavioral	HiBehavioral
	Ac look, watch	Ac stoop, wander	Ac find
	Mi perceive, observe	Mi bask?	
	Pa see, hear	Pa sit	Pa be angered
Biotic:	PrBiotic	KiBiotic	HiBiotic
	Ac live	Ac grow	Ac reproduce
	Pa receive nourishment	Pa be stimulated	Pa be bred, pollinated reproduced
Physical:	PrPhysical or Mathematical	KiPhysical or Kinematic	HiPhysical or Energetic
	Ac include, multiply, extend	Ac move (intransitive)	Ac affect
	Pa be included, be bounded, number.	Pa be moved	Pa be affected

(Functions are listed with Active (Ac), Middle (Mi), and Passive (Pa) examples belonging to the Functions. Hi=Priestly.)

3.13 *Order, interlocking (nonsovereignty) and luxuriance (nonreducibility) of Modes and Functions*

Now let us consider the relation of Modes and Functions to one another.

3.131 *W: order*

Is there any order of complexity, any order of higher and lower, among Modes and Functions? I think that I have implicitly answered this question in the course of the preceding discussion, and it remains only to make explicit what I have already done.

In the first place, there is an order of decreasing complexity among Modes: the order Personal, Behavioral, Biotic, Physical. This is already established by the order written into Genesis 1:28-30 of the corresponding Kingdoms: Human, Animal, Plant, Inorganic. Activities of the Personal Mode, generally speaking, require also motion (Physical), life (Biotic), and perception (Behavioral) on the part of the actor. Speaking, for example, requires motion of the mouth, direction of the speaking by a living Creature, and a kinesthetic and/ or auditory feedback to help regulate speech. Similarly, activities of the other Modes generally require characteristics from lower but not higher Modes. Animal behavior requires a *living* animal and one capable of some kind of *physical* response. It does *not* require speaking ability. The life of plants does *not* require speaking, ruling, jumping, swimming, etc.

In the second place, there is a decreasing order among three of the Functions: the order Sabbatical, Social, Laboratorial. This order, of course, is derivative from the order God, Man, Subhuman Creature related to the ordinances of sabbath, family, and labor. Nevertheless, the three creation ordinances are so bound together in the life of man that he cannot be involved in one of them without, in some fashion, being involved in the other two (whether for God or against him). For example, the Bible teaches that man *knows* God (Rom. 1:21)—even though "knowing" in general is classified under the Laboratorial Function.

On the one hand, man's loving a fellow man has "higher priority" than man's loving an animal; on the other hand, it cannot be said with the same confidence that man's *discussion* with his fellow man (Lingual) has "higher priority" than man's *studying* his fellow man (Cognitional). Or (to take the crucial case) is Christ's speaking to men of greater importance than his knowing men? In one sense yes, because if he knew us but would not speak to us, we could not be saved; in another sense, no, because he must know us to speak saving words to us. Hence one can speak of an order of importance only with caution.

Third, I surmise that there is even an order of Prophetic, Kingly, and Priestly. However, this can hardly be an order of importance. Can we say that it was more importance that Christ be prophet for us than that he be king and priest? Or priest than prophet and king? These are not choices at all. Moreover, it is difficult to see what sense could be made of the claim that one of these three is more complex than the others. If Prophetic activity requires Kingly (one must have power in order to speak), equally the Kingly requires the Prophetic (a king must be able to communicate his will).[6]

That there is an order of Prophetic, Kingly, Priestly can be seen in a number of ways. (a) This order expresses itself in the history of redemption (cf. 3.2). The Old Testament period or preparation period has an emphasis on the Prophetic, because it looks forward to a coming salvation (e.g., I Pet. 1:10-11; Rom. 15:4; Luke 24:27). The period of the accomplishment of redemption, in Christ's earthly life, has an emphasis on the Kingly ("the kingdom of God"), because it is the period of the decisive exertion of God's saving power (cf. Mark 1:15; 4:26, etc.). The period of the application of redemption, after Pentecost, has an emphasis on the Priestly (communion with Christ, "in Christ," the pouring out of the *blessing* of the Holy Spirit), because it is the period of the receiving of the benefits of divine fellowship. However, it is also obvious that Prophetic, Kingly, and Priestly activities are all present intensely in *each* of the periods. The transition from one to another is only a faint shift in emphasis.

(b) The order Prophetic, Kingly, Priestly expresses itself in the order of elements in covenant revelation of God, especially those biblical covenants related in form to Hittite suzerainty treaties.[7] The second millennium Hittite suzerainty treaties included (1) identification of the suzerain, (2) historical prologue (these two are Prophetic), (3) stipulations, (4) provision for covenant preservation and remembrance, (5) covenant witnesses (these three are basically legal, and therefore fall most clearly under Kingly), (6) curses and blessings (Priestly).[8]

(c) The order Prophetic, Kingly, Priestly seems to be derivative from the Trinitarian order of Father, Son, and Holy Spirit. One can argue for this directly, if one is willing to use a somewhat oversimplified, overschematic view of the Trinity. For example, Abraham Kuyper has it that the Father brings forth, the Son arranges, and the Holy Spirit perfects.[9] That kind of distinction is evidently related to the distinction among Prophetic, Kingly, and Priestly activity.

If, however, one hesitates to rely on an oversimplification, one still has the alternative of noticing that the role of the Father comes into prominence in the OT, the role of the Son in the Gospels, and the role of the Holy Spirit in the post-Pentecost application of redemption.[10] Thus there is at least an indirect correspondence between the persons of the Trinity and Prophetic, Kingly, and Priestly Functions. Now, just as there is no intrinsic subordinationism in the Trinity, so there is no intrinsic subordination among these three Functions.

3.132 F: interlocking (nonsovereignty)

What, now, is the relation among the Functions? I have already been discussing this, at least indirectly, all along. I will discuss it further in 3.2 and 3.3. But if I had to sum it up in one word, the word would be 'interlocking' (in which I include overlapping). This interlocking is obvious if one thinks of the Prophetic, Kingly, and Priestly Functions as they occur in Scripture. Especially the attempt to correlate these Functions with preparation, accomplishment, and application of redemption is full of interlocking and fuzziness. There is a marvelous, unfathomable richness in the biblical material on these Functions.

Hence I feel that a phrase like 'sphere sovereignty' or 'sovereignty of Functions' would be inappropriate here.[11] Functions are not "sovereign." In fact, I want to use the term 'interlocking' to indicate a kind of "nonsovereignty," if you will. OT kings are bound to submit to the word of an OT prophet, even when the prophet's word goes against the "good" judgments of statescraft (e.g., II Chron. 16:7-9; 18:1-34; 20:35-37; Jer. 27:8-15). The Levites (Priestly) receive an ordering imposed by David the king (I Chron. 23–26). The interaction among the three Functions ill comports with applying the word 'sovereignty' to each separately.

I naturally have no objection to speaking of the sovereignty of the Father, Son, and Holy Spirit as prophet, king, and priest.

3.133 P: luxuriance (nonreducibility)

Along with the interlocking of different Functions there is a luxuriant richness to each. The Functions differ from one another. We have already shown this, for most of the Functions, by simply observing the distinctions made in the Bible itself between Prophetic, Kingly, and Priestly, Sabbatical, Social, and Laboratorial, etc. Hence there should be no attempt to "reduce" one Mode or Function to another as non-Christian philosophies have sometimes attempted to do. For example, a materialist explains everything in terms of the Physical (life or personality is not regarded as anything special), the evolutionist in terms of the Biotic, the behaviorist in terms of the Behavioral, the Pythagorean in terms of the Mathematical, the Marxist in terms of the Economic, and so forth.

The basic problem of all such "reductionisms" is their idolatrous character. One Mode or Function is used as an ultimate explanation for Creation, and thus as a substitute for God. Instead of thanking God for each form of variety in his Creation, people suppose that they do not need to thank anyone.

But what is "reductionism"? "Reducing" everything to one Function (say the Energetic) could mean one of several things. First, it could mean simply preferring to describe everything in terms of Energetic characteristics, emphasizing Energetic characteristics, seeing everything in the context of the Energetic. Let us call this procedure "emphasizing reductionism."

Description. *Emphasizing Reductionism* is preference for or preoccupation with one (or a small number of) viewpoint(s), Mode(s), Function(s), or other Item(s) when one attempts to discuss and interpret some subject matter.

Now Emphasizing Reductionism could be *either* right or wrong, depending on whether or not there were idolatrous purposes involved. Certainly a physicist might prefer to look at things in terms of the Energetic Function, without being idolatrous about it. Moreover, in a certain sense everything *can* be explained in terms of the Energetic Function. God causes whatever happens, and that is an adequate explanation for everything.[12] Or everything could be explained in Economic categories: God is the *giver* of all (Acts 17:25). Or Lingually: God *speaks,* and it is done (Heb. 1:3). And so on. The idolatrous forms of reductionism are plausible precisely because of this ability to explain things in terms of any one of several Functions of God. If reductionisms are idolatrous, it is because of the way they go about their business—they leave out the true God.

Second, "reductionism" could mean refusing to acknowledge the legitimacy of other kinds of emphases and other kinds of language than one's own. For instance, the Energetic reductionist might complain to someone who talked about economics, "You *talk* that way, but *really* there's nothing to it but motions of molecules. What we call "money" is nothing but sheets of processed wood pulp with pigments on its surface. Etc." Let us call this type of argument "exclusive reductionism."

> Description. *Exclusive Reductionism* is the insistence on the exclusive correctness of one's own form of Emphasizing Reductionism.

Third, "reductionism" could mean the ambiguous use of key Functional terms in a broad sense and in a narrow sense, in order to construct a non-Christian "ultimate explanation" of the Cosmos. For example, suppose that an Energetic reductionist uses 'physical' both in a broad sense (like my 'Physical') and in a narrow sense (e.g., "physical" as dealing with the ultimate stuff of modern physics). Then at one point in his discussion *everything* is subsumed under the "physical"—it is a *universal* explanation—and at another point the "physical" denotes modern physics that does the explaining—it is a universal *explanation*. A similar semantic sleight of hand takes place if a materialist alternately uses 'matter' to mean "*everything* that exists" and "the underlying stuff of *nonliving* things." Let us call this type of position "slippery reductionism."

> Description. *Slippery Reductionism* is the ambiguous use of key terms in a broad sense and in a narrow sense, in order to construct a non-Christian "ultimate explanation" of the Cosmos.

Needless to say, combinations of these Reductionisms are possible. Exclusive and Slippery Reductionisms use Emphasizing Reductionism in order to sound convincing.

The problem of Reductionism has been one of the main motivations for introducing so many technical terms like 'Prophetic,' 'Behavioral,' and 'Cosmos.' I shall introduce many more in the following sections. It is to help us to be able to use many different perspectives and emphases as the need arises. We need both to avoid ourselves Exclusive and Slippery Reductionisms, and to be alert to detect such reductionisms in others and to know how to deal with them. Having a big technical vocabulary is not important "for its own sake," but for the sake of having a large number of convenient tools for dealing with reductionisms.

Though the examples in this section are "Functionalistic" reductionisms, that is, reductionisms engaged in "reducing" one Function or Mode to another, similar types of "reduction" can occur in other of the areas of "temporality," "axiology," etc., that I shall describe in the following sections and chapters.

In the most general sense, this book is itself a form of Emphasizing Reductionism. I am emphasizing certain things, and leaving other things out. But I have tried to guard at least partially against Exclusive Reductionism by pointing out that people could use other technical terms than mine, and could make distinctions in other ways than as I have made them. I have tried to guard against Slippery Reductionism by distinguishing my technical terms from ordinary language by the device of capitalization.

Now, what are we to say by way of criticism of Reductionisms? Obviously, Emphasizing Reductionism is not wrong unless it is mixed with Exclusive or Slippery Reductionism. But frequently we do have to deal with a mixture. I suggest three complementary ways of confronting bad cases of Reductionism. We can use (1) a more or less direct confrontation with the Bible (ontological criticism: 3.1331), (2) an analysis of faulty methodology in the Reductionism (methodological criticism: 3.1332), and (3) critique of the Reductionism's self-validation (axiological criticism 3.1333).

3.1331 Ontological criticism

One kind of criticism involves direct confrontation between a false reductionism and a true account.[13] This confrontation can take place in at least three ways. Way 1 appeals directly to the language of the Bible. The Bible uses a great variety of language about all Functions. Such language is not erroneous. Hence there is nothing sacrosanct about thinking and talking in terms of only *one* Function. This contradicts Exclusive Reductionism. Slippery Reductionism can be counteracted by observing that in Scripture God himself serves as "ultimate explanation." Using another explanation to escape God is idolatry.

Way 2 involves a more direct appeal to the fact that God himself is active in each of the Functions. God lives (Biotic), sees (Behavioral), speaks (Lingual), and so on (see 3.11, 3.1211). So a reduction of (e.g.) the Lingual to the Behavioral would include a reduction of *God's* speech to God's Behavior, a denial of the incomprehensibility of God.

Or, to put it another way, reductionism in effect denies the *perfection* of God. For God has in fact chosen to speak about himself in Scripture in a great variety of ways. To claim that some *one* Function is somehow "intrinsically" best for description (and hence in particular for the description of God) would involve an attempt to improve the language of the Bible, and hence to "correct" God. It also involves a claim to deeper knowledge of God than the Bible is able to give. Here the denial of incomprehensibility re-emerges.

Way 3 of confrontation is to "agree" with the reductionist—while at the same time redefining the words involved. For example, all *can* be explained Economically, *provided that* it is understood that

God is the giver and the ultimate value (*summum bonum*). Then one redefines Economic value in terms of God's rewards instead of merely in terms of food and drink (cf. Luke 12:22). The Bible tells us what it *really* means to be properly Economically oriented; namely, to participate in Priestly, Personal communion of blessing with God the eternal and consummate value.

Similar "redefinitions" can occur with other Functions.

3.1332 *Methodological criticism*

Another way of criticizing reductionism is by observing in a more internal way that the proposed reduction does not really "reduce."[14] Naturally, the criticism must proceed in somewhat different fashion depending on whether the Reductionism in question is Exclusive or Slippery or both. Now Exclusive Reductionism can take at least two forms. The reductionist can say that talk about some Functions "really" means talk about others (e.g., Economics is "really" part of Energetics). Or he can say that talk about some Functions is meaningless. Take as an example of the first kind the idea that Economics talk is "really" Energetic. This kind of claim need not be challenged "head on." Suppose it is true. Presumably one could go on using whatever Economic language one liked, so long as he bore in mind the "translation value" of the Economic language into Energetics. If Economic talk is just shorthand for Energetic talk—still, why not use shorthand? And so everything is left just as it was before.

Then just what is the value of pointing out that Economics is "really" Energetics? Perhaps it is claimed that Energetic language is more precise, or less likely to breed confusion and error. The answer is, it is more precise for some purposes, like the physicist's purposes. Moreover, why should precision be at a premium? Sometimes precision is simply pedantic. So the Exclusive reductionist is likely to be forced back into an Emphasizing Reductionism: he has a personal preference for Energetic language. Or else he will make some judgments about how we "ought" to speak, though he has no grounds for the "ought" other than his personal preference.

Suppose next that the Exclusive reductionist takes the route of denying that (e.g.) Economic language is meaningful. In that case, he is involved also in a Slippery Reductionism, since he is using 'meaningful' ambiguously in the broad sense of ordinary English (call it "meaningful$_1$") and in the narrow sense of "meaningful" language = Energetic language (call it "meaningful$_2$"). If he sticks exclusively to *one* of these two meanings of 'meaningful,' he cannot easily find an argument against Economic language.

If he says, for example, that Economic language is not meaningful$_2$, we agree. This is so by definition of 'meaningful$_2$.' If he says that Economic language is not meaningful$_1$, we disagree. When someone says, "I'll give you $20 for the lawn-mower," doesen't everyone understand what he means, including our reductionist? If he does not understand an utterance like that, he cannot function in human· society. So Economic language is meaningful$_1$ ("meaningful" in the sense of ordinary English).

Logical positivism is an example of a Lingual Slippery Reductionism, based on similar play on the word 'meaningful.' Suppose we call "meaningful$_3$," any language approved by the positivist. Then other language is not meaningful$_3$, but why not go on using it to accomplish our purposes, to talk about God, metaphysics, and so on? Such talk is still not necessarily in the same talk as baby-talk or gibberish. The logical positivist succeeds only by surreptitiously importing into 'meaningful$_3$, the positive connotations of ordinary English 'meaningful$_1$.' If something is not "meaningful," we need not attend to it. 'Cognatively meaningful,' a phrase also used in this connection, is beset with similar slipperiness.

How do we deal with Slippery Reductionism (reductionism using ambiguous key terms)? The obvious way of proceeding is to ask for some specification of how ambiguous terms will be used, and then hold the reductionist to his proposed usage. Or, lacking such a specification, one can trace through the ambiguities of usage. This is harder than it sounds, because the ambiguities are usually very subtle.[15]

3.1333 Axiological criticism

A third way of criticism is open, namely of asking whether a supposed reductionism offers adequate grounds for its own validity. Naturally, these three ways of criticism (3.1331, 3.1332, 3.1333) are complementary, interlocking, and mutually reinforcing, rather than being completely separate.

First, one can ask whether Reductionism does not remove the ladder used to climb up to its own conclusions.[16] For example, against the materialist one points out that the Physical observations used to arrive at the materialist theory were made by people who naïvely assumed that they could draw an operable distinction between themselves and animals, plants, and nonliving things. Since their naïve assumptions were wrong, the validity of their observations is called in question, and the observational support for the materialist theory collapses. C. S. Lewis has constructed a similar refutation of evolutionary naturalism.[17] Another example: Exclusive Reductionism of the Lingual to the Economic is undermined by the observation that construction of the Economic theory depended on faith in the basic meaningfulness of language. And so on.

Secondly, one can ask whether Reductionism abolishes human knowing in any meaningful sense. If so, with that abolition it also abolishes knowledge of the truth of Reductionism. If, for example, everything is reduced to the Economic, then *knowledge* of the Economic is also reduced, and the result may be "knowing" determined by Economic factors. Does this discredit one's *own* supposed knowledge? This is a problem to a Marxist who wants to discredit *other* people's ideas by appealing to the fact that those ideas are determined by their relation to the means of production. Why aren't his own ideas equally discredited?

Thirdly, one can ask whether Reductionism allows room for any purpose for its own existence. If, with the materialist, one is convinced that all is the motion of matter, surely it can't matter much whether the "matter" believes in the materialist theory. Human life is ultimately purposeless.

3.2 *Temporality*

We have now come to the second of the three sections sketched in the beginning remarks of chapter 3: discussion of the historical development of the Creation under the direction of Men.

In discussing this development, we shall be constructing an outline of a philosophy of history. This is important not only for its own sake, but for the purpose of obtaining a proper perspective on the scientific task. What we are to do, what goals we are to have, depends to some extent (but how much? and in what ways?) on what period of history we are in. OT food regulations no longer have the same relation to us as they did to the people in OT times. Nor will scientific work have the same role after the return of Christ as it does now.

Another question: are so-called scientific "laws" valid for the past with the same degree of accuracy as they are now? Will they be valid in the future? Answering these questions involves taking into account what kind of changes are involved in the fall of man into sin, in Noah's flood, and in the return of Christ in the future.

Once again, we can begin with God's instructions in Genesis 1: 28-30. The second set of distinctions in Table 1 is of particular relevance. Man is to have dominion over the animals, to eat the plants, and to fill and subdue the earth. Here is a program for change, for development, which might involve several phases.

3.21 *Major divisions of periods*

Adam, of course, failed to keep God's covenant, and so we do not know what would have happened apart from the fall. Christ, however is the last Adam, as we have seen (3.1212). It is in terms of Christ's work that we must understand the course that history has taken. The most obvious division of history is thus into (a) the period preceding the coming of Christ, (b) the period of Christ's appearing in the flesh, and (c) the period of fruits following Christ's appearing.

> Description. The temporal development of Creation can be divided into the *Preparation Period,* the *Accomplishment Period,* and the *Application Period* of redemption, that is, the OT period, the period of Jesus birth, life, death, resurrection, and ascension (the Gospels), and the period of the application of the benefits that he has won (Acts and onwards).

Once again, my language should not be interpreted rigidly. Did John the Baptist belong to the Preparation or to the Accomplishment?

One could make a plausible case for either one, in view of Luke 7: 26-28. And I certainly do not intend to say that redemption was not applied in the OT (in that case, no OT saints would have been saved). I have used the term 'Application' to refer to the period of application *par excellence.*

The threefold division into Preparation, Accomplishment, and Application is so obvious that it is somewhat artificial to call on proof texts. However, three stages are visible in Galatians 4:3-7, Luke 24:44-47, Acts 13:17-41 (vv. 17-22 Preparation, 23-27 Accomplishment, 38-41 Application; note, however, hints of Application sprinkled in vv. 23-27).

3.22 The Accomplishment Period

These Periods can themselves be subdivided. For example, Scripture instructs us to understand Christ's work in terms of the sequence of suffering to glory (Luke 24:26, 46; Acts 2:22-24, etc.). In a way, the sufferings and glory of Christ are a kind of "small-scale" Accomplishment and Application, since as a result of his obedience terminating in death (Accomplishment: Heb. 5:8; Phil. 2:8), he receives in his own person the fruits and reward of service (Application: John 17:5). The "Preparation" for his sufferings is then found in the ministry of John the Baptist, and Christ's birth, boyhood, and baptism.

Of course, it would be a mistake to search for perfectly detailed parallelism between the life of Christ and the Preparation, Accomplishment, and Application Periods. Hence I will use different terminology.

> Description. The *Generational, Developmental,* and *Culminational Accomplishment Periods* are, respectively, the periods comprising (1) the birth narratives to the baptism of Jesus, (2) the temptation to Gethsemane, and (3) the trial to the resurrection, ascension, and enthronement.

You may ask why I did not start the Culminational Accomplishment Period with the resurrection of Christ. Everyone would agree that the resurrection is *the* great event in the glorification of Christ. I include the crucifixion with the third period (a) partly because the crucifixion of Christ can already be viewed as the beginning of his glorification. In John 3:14, 8:28, and 12:32 the "lifting up" includes both crucifixion and exaltation, as is clear especially from the serpent-pole imagery in 3:14. I do it also (b) because I want to retain the close link between crucifixion and resurrection so frequently expressed in the NT (Rom. 4:25; 6:4; Acts 2:36). (c) Moreover, as we shall see below (3.25), my classification of Periods better exhibits the relation of the periods to the Prophetic, Kingly, and Priestly Functions.

3.23 The Application Period

Note next that the pattern of suffering and glory in Christ's life is mirrored in the believer's life: "I consider that the sufferings of this present time are not worth comparing with the glory that is to be revealed to us" (Rom. 8:18). "For while we live we are always being given up to death for Jesus' sake, so that the life of Jesus may be manifested in our mortal flesh. . . . knowing that he who raised the Lord Jesus will raise us also with Jesus and bring us with you into his presence" (II Cor. 4:11, 14). "That I . . . may share his sufferings, becoming like him in his death, that if possible I may attain the resurrection from the dead" (Phil. 3:10-11). See I Peter 2:21. Of course, there are also points of difference. The believer's sufferings do not have atoning value, and he even *now* shares in the resurrection life of Jesus Christ (II Cor. 4:11; Phil. 3:10; Col. 3:1-4). One must beware of overschematizing.

But having issued the cautions, I proceed to draw out some of the similarity terminologically.

> Description. The *Individual Generational, Developmental,* and *Culminational Application Periods* of the Christian's redemption are, respectively, (1) the time of conversion, of initiation into God's people; (2) the period of walking with Christ in this world; and (3) the period of glory initiated by death or the return of Christ.

Of course, one could also apply this language of stages to the corporate life of the church.

> Description. The *Corporate* Generational, Developmental and

Culminational Application Periods are, respectively, (1) the time of Pentecost and the founding of the church (Acts), (2) the history of the church from its founding to the coming of Christ (Revelation), and (3) the glorification of the church at the coming of Christ, and the time to follow.

3.24 The Preparation Period

The next step would logically be to see whether the division into Generational, Developmental, and Culminational Periods can legitimately be applied to the OT. But if such Periods exist in the OT, the dividing lines are much fuzzier than in the NT, and so the proposed division loses some plausibility.

However, it cannot be said that OT history is one unbroken continuum, with no shifts in pattern. The OT itself groups some sequences of events together, and the NT pattern of interpretation of the OT is also sensitive to grouping (cf., e.g., the condensed OT histories in Acts 7:2-53; 13:17-41). The OT groups its events around at least three centers: (a) prominent individuals like the patriarchs Noah, Abraham, Isaac, Jacob, and Joseph; the leaders Moses and Joshua; the kings Saul, David, and Solomon; and the restoration leaders Zerubbabel, Joshua, Ezra, and Nehemiah; (b) the great epochs of redemption such as the flood, the birth of Isaac, the Exodus, the conquest, the triumph of David and Solomon's kingdom, and the restoration from exile; (c) the covenants connected with individuals and the redemptive epoch: Noachic (Gen. 9:1-17), Abrahamic (Gen. 12:1-3; 13:14-17; 17:1-27; 35:9-15), Mosaic (Exod. 20 onwards, but especially 20–24), Davidic (II Sam. 7:4-17), Solomonic (I Kings 9:1-9; II Chron. 7:12-22), and restorational (Hag. 2:2-9).

There can be no doubt, therefore, that the OT period can rightly be divided into some 5-10 major epochs, provided that one is not overly concerned about fixing exact boundaries to the epochs. The remaining question is this: which epochs and which prominent persons are to be subgrouped with which? A little reflection shows that this is not a question that one would have to answer one way to the exclusion of another. Any one of several subgroupings is all right, provided that it does not claim to exclude the elements of continuity that an alternative subgrouping makes explicit.

In some ways, for example, Joshua can be grouped with Moses, because he is continuing the program of deliverance into the promised land set in motion through Moses (cf. Exod. 3:17). Or the Book of Joshua could be grouped with Judges and the following books, because all deal with situations in the promised land. If Joshua is a second Moses with the spirit of wisdom (Deut. 34:9; Josh. 1:5), he is also related (though in a vaguer way) to the judges who follow him.

On the other hand, some subgroupings are perhaps more significant than others. For example, Abraham is linked more closely to Isaac than to Noah. Joshua belongs, I think, more with the Mosaic period than with the period starting with the judges and continuing through the kings of Israel. The emphasis of the Book of Joshua is more on ending something by a fulfillment of promises already given (Josh. 11:15-23; 23:14; 24:2-13), than on starting something with a series of promises not yet fulfilled. It is a period, if you will, of resolving tension rather than building it (in contrast to Judges). Nevertheless, the two-sidedness is shown by the fact that there is *some* concern for the future, particularly in the later chapters 23 and 24 (23:5-13, 15-16; 24:14-15).

If, now, one wants to group together all the more closely related individuals, redemptive epochs, and covenants, one obvious major grouping is a grouping into the patriarchal period (Genesis), the Mosaic period (Exodus–Joshua), and the Davidic period (or period of kings: Judges–Nehemiah). Each of these periods has its own flavor. The patriarchal period is a background period, somewhat amorphous, characterized by intensification of the promise of seed and land. The Mosaic period is a period of law-giving. The Davidic period is a period of the kingship.[18]

But then, on second thought, the patriarchal period is a kind of preface to the Mosaic period. It was written by Moses as a backdrop for Exodus–Joshua, to explain the origin, nature, and purpose of the nation of Israel in the midst of the nations. Hence one could see the whole of OT history as composed of only two periods, Mosaic and Davidic.

A second problem is in fitting the restoration into the Davidic period, because kingship does not play such an obvious role in exilic and
postexilic times as it does in Judges–II Chronicles. Thus one may
divide the Davidic period into a period of establishment of kingship
and kingdom (Judges–I Kings 7), and a "sanction" period where the
primary concern is on the covenant sanctions[19]—whether God will be
with his people in blessing, or bring the curses of the covenant on
their heads (cf. the concern of II Chron. 7:12-22). The exile is
then a kind of "curse" stage, and the restoration a "blessing" stage.

Which of these proposed divisions of the OT is "right"? Why,
any one of them, of course. All of them serve to point out aspects
of the continuity and movement in OT history. In order not to clutter
things up, however, I will stick with one of them, that one which I
think most easily illustrates a continuity with the Generational, Developmental, and Culminational Periods in the NT.

> Description. The *Generational, Developmental,* and *Culmi-*
> *national Preparation Periods* are, respectively, the Mosaic period
> (roughly Gen. 3:8–Joshua), the period of establishment of king-
> ship and kingdom (roughly Judges–I Kings 7 or 10), and the
> period of execution of sanctions (roughly I Kings 8 or 11 to
> Nehemiah).[20]

3.25 Interlocking

One can now see an interesting interlocking between the Prophetic, Kingly, and Priestly Functions on the one hand and the Generational, Developmental, and Culminational Periods on the other.
Not as if only one of these Functions were operative in the periods
in question. But in a Generational Period the Prophetic Function
attains a prominence greater than it does in the corresponding Developmental and Culminational Periods. In a Developmental Period the Kingly Function attains prominence. And likewise, in a
Culminational Period, the Priestly Function.

The clearest demonstration of this is in the life of Christ. At the
beginning of his ministry stands a *call* from God (Luke 4:18-19;
Heb. 5:4, and his baptism by John [Mark 1:11]; indeed John himself
is merely "a voice" [John 1:23]); in the middle stands his proclama-
tion of the *kingdom* of God of which he is the embodiment (John

5:17; 9:4); at the end stands his *priestly* offering of himself to his Father and his present intercession (Heb. 9:14; 7:25).

One can also look at the Culminational Period from the standpoint of the Father's work. The Culminational Accomplishment Period is the Period of execution of the covenant sanctions on Christ, both curses and blessings. This belongs to the Priestly Function of the Father.

One can see a similar pattern of Prophetic, Kingly, and Priestly work in the Applicational Period. The beginning of a believer's new life can be summarized in the word 'call' (Eph. 4:1; I Cor. 1:9), the course of his life as a "walk" in which God works in him (Phil. 2:13; Gal. 2:20), and the end as a reward (Phil. 3:14; II Tim. 4:8) or "glory" (Rom. 8:18, 30, etc.). These three clearly correlate (with some slippage, yes) with what I have earlier called the Prophetic ("calling"), Kingly ("working"), and Priestly ("glorifying") Functions of God.

Finally, let us look at the three sub-Periods of the Preparation Period. If we look at the three Periods (Mosaic, "Davidic," and "sanctional") from the standpoint of the human *offices* of prophet, king, and priest, then the order appears to be reversed. For example, in the Mosaic Period (Generational Preparation Period) far more attention is devoted (in terms of quantity of material) to the priesthood and its duties than to either the prophet or the king (but cf. Deut. 17–18). The Developmental Preparation Period is, of course, the period of development of the kingly office (Judges shows the poor results when the people have leaders that are not yet kings; see the refrain of Judges 17:6; 19:1; 21:25). The Culminational Preparation Period sees a blossoming of the office of the prophet (beginning with I Kings 13:2ff., and climaxing with the great writing prophets).

On the other hand, if we look at the process more from the standpoint of what *God* does than from the standpoint of what offices men fill, a different result emerges. The Mosaic Period is above all a period of law-giving, a period therefore of God's speaking. It cannot be said that either of the two subsequent Periods adds substantially to Mosaic law. The prophets, rather, call people back to the law, and warn what the results will be if their disobedience continues. Second, the Davidic Period is the great period for the establishment of God's kingdom. Thirdly, the third or Culminational Preparation Period is the time for God's pouring out of his covenant blessings and cursings, already Prophetically pronounced in (e.g.) Deuteronomy 27–30. That is not to say that blessing and curse were *not* operative at an earlier stage; for example, in the wilderness (or even in the time of Noah!). It is only to say that the focus is, in the Culminational Period, more consistently on this area.

The Balance is also redressed by the observation that, if we confine ourselves to the historical accounts and the chief human officers in these accounts, the result is a pretty clear order of prophet, king, priest. In the Generational Preparation Period the chief officer is Moses himself—the principal reason why there are not many prophets on the scene. He is a prophet but more than a prophet (Num. 12: 6-8; Deut. 18:15). In the Developmental Preparation Period the chief officers are judges and kings. In the Culminational Preparation Period, the historical accounts of I & II Kings do emphasize the kings and prophets, but Chronicles (consistently with the later date of authorship) is more interested in the priests and Levites. Priests retain importance in Ezra–Nehemiah.

I could, of course, proceed to introduce further subdivisions in the Periods that I have already described. A greater amount of arbitrariness, however, would be unavoidable in such a subdivision. This is true not only because the lines of subdivision are less clearly marked than for major divisions, but also because of the two-way connections. All periods have connections both forward and backward. Any period of biblical history can be regarded from the standpoint either of (a) what it presages and prepares for and forms the background of in the future, or of (b) what is accomplished in the period itself or of (c) what it is an outgrowth and fulfillment and recompense for in the past.

I grant that some periods are more forward looking (the account of Abraham looks forward to Isaac and Jacob more than it looks back to Noah and Enoch), and some are more backward looking (I have argued that Joshua is). Yet in the nature of the case for the historical accounts in the Bible, both forward and backward references are detectable all the way along. God's program must move from inception to conclusion. The result is that *any* passage can be viewed as Generational, Developmental, or Culminational.

Description. The *Generational, Developmental,* and *Culminational Views* are, respectively, those views of an event or complex of events which see the events primarily in terms of (a) presaging a future, (b) accomplishing certain tasks within the locus of events in question (c) fulfilling the past behind the events in question.

For example, the Generational View of the story of Abraham would see it in terms of the intensification and specification of the promise of a "seed" Christ through whom all the families of the earth would be blessed (Gal. 3:16; Rom. 4:16-17; Ps. 72:17). The Developmental View would see it in terms of the formation of a separate people of God among the nations, a people living by the promise of God and marked by sacramental seal (Gen. 17). The Culminational View would see it as God's response to the sinful plans of men in Babel. They wished to build to heaven, to make a name for themselves, and to appropriate for themselves a fixed piece of land (Gen. 11:4-5). After God judges them (11:6-9), he reverses the pattern by coming *down* from heaven to Abraham, by giving Abraham a name, and by giving him a land in which he will be a sojourner in his own life (Gen. 12:1-2).

Applying the Generational, Developmental, and Culminational Views to the Developmental Application Period has some interesting results. The Developmental View sees the present Christian life as work and progress, as a "walk." The Culminational View sees the Christian life in terms of participation in the benefits of Christ's work, a having entered into the new age—thus an "already" (Col. 1:13). The Generational View reminds us that we yet look forward to final salvation—thus a "not yet" (Rom. 8:18-25).

NT biblical theology is fond not only of using the terms 'already' and 'not yet,' but of pointing to the supposed dialectical tension and paradox between the two ideas.[21] Doubtless there is such a thing as groaning in the Christian life (Rom. 8:22-23; though this is characteristic more of the Generational View). But my terminological innovations will have done a service if they help show (a) that 'tension' is perhaps a less appropriate word than 'richness of perspective'; (b) that "already and "not yet" (in the form of Culminational and Generational Views) are already characteristic of the OT; (c) that there is a third view, the Developmental View, that speaks of growth, of progress, of working, serving thereby to round out and balance the Generational and Culminational Views.

It should now be clearer why there is a correlation between Prophetic, Kingly, Priestly, and Generational, Developmental, Culminational. By and large, the Generational view brings into prominence what a situation "tells us" of the future; the Developmental View focuses on the action, thus highlighting the Kingly; and the Culminational View sees a situation as reward or punishment, blessing or curse, brought by God in view of the past. That does not mean that we have here a one-to-one correspondence, but rather an interlocking.

3.26 Adamic history

One period of history has not yet been covered by our discussion: the period of Genesis 1:1–3:7. Of course, for some purposes, this might simply be included under the Generational Preparation Period. However, in some ways Genesis 1:1–3:7 is radically different, because it does not have the fallen situation of man already as a background.

One way of seeing elements of continuity between Genesis 1:1–3:7 and the rest of Scripture is by developing the relevant material in Romans 5:12-21 and I Corinthians 15:44ff. Adam and Christ are both representative figures whose actions draw their offspring (those "in" them) after them. Where Adam failed to withstand the tempter, Christ withstood (Matt. 4:1-11). Hence a certain amount of parallelism can be expected between their histories.

The parallel is not hard to see.

Description. The *Adamic Preparation Period* is the period of creation (Gen. 1:1–2:3). The *Adamic Accomplishment Period* is the period of Adam's probation (Gen. 2:4–3:7). The *Adamic*

Application Period is the period from Genesis 3:8 onward to the end of time.

As in the case of Christ, so in the case of Adam the Application Period is the time for applying the covenant sanctions consequent on the achievements of the covenant head (Rom. 5:12-21).[22] These Periods can in turn be subdivided in terms of Generational, Developmental, and Culminational categories (though one may not be so convinced of the nonartificial character of these).

Description. The *Generational* Adamic Preparation Period is the initiation of creation in Genesis 1:1-2. The *Developmental* Adamic Preparation Period is the development of creation in Genesis 1:3-31. The *Culminational* Adamic Preparation Period is the completion of creation in Genesis 2:1-3.[23]

Description. Likewise the *Generational, Developmental,* and *Culminational* Adamic Accomplishment Periods are constituted, respectively, by the call of Adam (Gen. 1:28-30; 2:16-17), the work of God in Adam (2:20, 23, 25, plus some not recorded), and the fall of Adam (3:6-7).

Description. The *Corporate Generational, Developmental,* and *Culminational* Adamic Application Periods are constituted, respectively, by the giving of the curse (Gen. 3:8ff.), the development of the wickedness of this world (Gen. 4:1ff.), and the punishment of the wicked at the last judgment (Matt. 25:41-46).

Since both Adam and Christ are federal representatives, the Adamic and Christological "Periods" overlap.

3.27 *Vocative, Dynamic, Appraisive*

Next, we can stress in another way the interlocking of Prophetic, Kingly, and Priestly Functions with the history of redemption. Any one Period or part of a Period can be analyzed in terms of what God is doing in a Prophetic, Kingly, or Priestly manner. This may involve either analyzing *one* act of God in terms of the three Functions, or it may involve showing how a certain bundle of acts of God brings a given Function into prominence.

Description. The *Vocative, Dynamic,* and *Appraisive* aspects of events are those aspects involving the Prophetic, Kingly, and Priestly Functions, respectively.[24]

My meaning can probably better be discerned from examples than from this rather vague description.

My first example is from the Developmental Adamic Preparation Period (Gen. 1:3-31). Vocative creation consists in the "let there be" commands (Gen. 1:3, 6, 9, 11, 14, 20, 24, 26). Dynamic creation consists in the formation of Creatures (Gen. 1:[4], 7, [9b], [11b], 16-17, 21, 25, 27, [31a]). Appraisive creation consists in the approbational sanction, "It is good" (1:5, 8, 10, 22, 28-30).

As a second example, take the redemption from Egypt. There is a Vocative redemption in the call of Moses at the burning bush, and the later commands to Pharaoh; a Dynamic redemption in the actual going out of Moses with the people from Egypt; and an Appraisive redemption in the meeting with God at Mt. Sinai.

Third, take the resurrection of Christ. A Vocative resurrection, or call to rise from the Father, can be inferred from Romans 4:17; the Dynamic resurrection is the actual raising of the Son by the Father; and the Appraisive resurrection is in Christ's being designated and declared Son of God in the resurrection (Rom. 1:4).

Fourth, take the Individual Generational Application Period, that is, the time of conversion of a believer. Vocative generation is what theologians have called "calling" (I Pet. 2:9); Dynamic generation is what they have called "regeneration" in the narrow sense (John 1:13; I John 5:1, etc.); and Appraisive generation includes justification (justification is the application of a judicial covenant sanction: Rom. 4). It should go without saying that we are not speaking here of three chronologically separated events, but three perspectives on the same event of being vitally united to Christ, of being created anew (II Cor. 5:17).

Similarly, the Culminational Application Period includes a Vocative consummation in the last trump, a Dynamic consummation in the resurrection of the dead, and an Appraisive consummation in the last judgment.

3.28 *Some preliminary implications*

An appreciation for the differences between the Periods sketched above will modify the conclusions of historians, of ethicists, and in some ways even of natural scientists. Extrapolations forward and backward in time, based on the present situation, should take into account these alterations between Periods.

The natural scientist, for example, cannot claim that his extrapolations are necessarily valid after the second coming or before the seventh day of creation (see further 5.3321). He has to reckon with the difference that God's announced policy for Israel (Deut. 28–30) may make in scientific description of Near Eastern agriculture, meteorology, and so forth, in the Developmental and Culminational Preparation Periods.

Again, the ethicist ought not to claim, as has sometimes been done, that we in the twentieth century have "outgrown" NT ethics in the same way that the NT outgrew OT ethics. A decisive transition occurred with the onset of the Accomplishment and Application Periods, the like of which has not occurred during the *course* of the Corporate Developmental Application Period, no matter how long that Period may have extended by the clock. Conversely, NT ethics may not be used to undermine or challenge the divine origin of (say) the OT sacrificial system, as if this represented an "inferior" or "primitive" stage of religion. The OT patterns of worship were designed by God to meet the needs of the Preparation Period. It is no wonder that they are unsuited (at least in external details) to the Application Period.

3.3 *Structurality*

Structurality is the third of the three topics outlined in the beginning of chapter 3. In accordance with part "F" of Table 1, I now discuss "how things function" in terms of connections and relations. In Table 1, the connections and relations are connections and relations among the four Creational Kingdoms. However, this section discusses not only those relations, but also relations within a given Kingdom, relations of God to Creation, and relations of God to God.

An examination of such relations can help us to evaluate (a) the all-important role of "law" in scientific work (see 3.324) and (b) the relation of scientific organizations to other organizations of society.

3.31 *The Lord*

God has relations with himself. The Father loves the Son (John

3:35); the Father knows the Son and the Son knows the Father (Matt. 11:27); the Spirit searches the depths of God (I Cor. 2: 10-11). The persons of the Trinity dwell mutually in one another (John 14:11, 23). But there is a great mystery to these relations. So let us pass on to consider God's relations to Creation, when and after he has created the Creation.

3.32 *The Bond*

First we consider the covenantal relation between God and his people.

3.321 *The Covenantal Bond*

Scholars have already argued at some length that "covenant" is a comprehensive category for interpreting and providing a framework for God's relations to man.[25] Though, as we shall see, God's words with respect to Creation in certain respects go beyond "covenant" (3.322), yet the biblical material on covenants can certainly form our starting point.

Now, (a) we could start from the word *b*^{e}*rit* (covenant) and the broad meaning that it has both in Scripture and in the ancient Near East in general. Or (b) we could start from the *particular* covenant that God ratifies and establishes with his people. Or (c) we could start from the diversity of covenants that God makes with Adam,[26] Noah, Abraham, Moses and those with him, Joshua and those with him, and so forth. These three —(a), (b), and (c)—are not the same. Let us spell out some of the differences.

(a) *B*^{e}*rit* means, roughly, pact or alliance under sanctions, whether between God and man (Gen. 15:18), between ruler and subjects (Jer. 34:8-11), or between equals (Gen. 21:27), whether between two individuals (Gen. 21:27; I Sam. 18:3), between one individual and a ruler with his subjects (II Kings 11:17), or between more than two parties (Gen. 14:13; Ps. 83:5[6]). (b) The particular covenant that God makes can be summed up in the words, "I will be your God, and you shall be my people" (Exod. 6:7; 19:5-6; Zech. 8:8; Lev. 26:12; II Cor. 6:16; Rev. 21:3). (c) The covenants made with Adam, Noah, Abraham, etc., obviously have different content, because of the differences in circumstances; however, there remains the underlying continuity of covenant in sense (b).[27]

Any one of these three senses could form the starting point for discussion. But (a) and (b) are easier to deal with because they offer us a *general* pattern at the beginning of our work. (b) might seem to be the most appropriate starting point, since there is nothing especially "sacred" about Near Eastern covenants in general. However, it is also true that, because man is the image of God, the covenants that men make with one another have some resemblance to the divine covenant. Therefore it matters not so much where we start.

Description. The *Covenantal Bond* is the pact under sanctions, revealed in Scripture, between God and his people. Scripture sums up the Covenantal Bond in the words, "I will be your God, and you shall be my people." The Covenantal Bond includes both law, administration, and sanctions of covenants.

Though this description of the Covenantal Bond has singled out the people of God, the rest of Creation is not *un*affected by the Covenantal Bond. The boundaries here are somewhat fluid. To begin with, not all those who receive the initiatory sign of the Covenantal Bond (circumcision in the OT, baptism in the NT) are finally saved, and yet the reception of the sign has an effect on them: their responsibility and guilt are thereby increased.

Secondly, the promises of the Covenantal Bond, at least as they become deepened in the course of redemptive history, include the promise of new heavens and a new earth (Isa. 65:17; II Pet. 3:13, etc.), implying a comprehensive renovation of all Creation.

Third, the Covenantal Bond has an explicit bearing on the reprobate and animals. The covenant of Gensis 9:8-17 is said to be between God and all Noah's descendants and every living creature (animals). We must maintain that this covenant has not been abrogated up till today. For one thing, no explicit abrogation of it occurs elsewhere in Scripture (see, in fact, Isa. 54:9; II Pet. 3:5, 7). Secondly, the sign of the covenant continues to be a guarantee of the promise (Gen. 9:13-17). Since the Noachic covenant includes all men, it also includes the reprobate. Furthermore, the reprobate are bound to the moral law (Rom. 2:14-15; 1:32), which law in the form of the ten commandments is covenant law (Exod. 34:27-28; cf. Deut. 4:13).

An objector might, of course, argue that the Noachic covenant has nothing to do with the Covenantal Bond, since it is with all men rather than specifically with the people of God. Now everyone must admit that the Noachic covenant has this distinctive function. But it is equally true that Genesis pictures the covenant as a gracious response on God's part to the sweet smelling sacrifice (8:20-22). And the OT sacrifices prefigure the final sacrifice of Christ. Hence the Noachic covenant should be viewed as one of the provisions, for the sake of Christ, in terms of which God undertakes to maintain Creation until the invitation of salvation goes to all nations and all of his people find salvation in Christ (II Pet. 3:9; John 6:39).

Scripture also shows that the relation of God to plants and Inorganic Creatures is bound up with the Covenantal Bond. But plants and Inorganic Creatures are relatively in the background. This is not surprising, since the major focus of the historical covenants in Scripture is on God's relation to men. God includes animals in the Noachic covenant because their lives, too, had been threatened by the flood and preserved through the ark. The flood did not pose the same danger to plants and Inorganic Creatures, so there was no motive for bringing them in as parties of the covenant. Nevertheless, they are mentioned indirectly in 9:1 and 9:3, in a fashion similar to what had been the case in the Adamic covenant (Gen. 1:28-30). Moreover, the promise of the covenant definitely involves certain commitments respecting Inorganic Creatures (Gen. 8:22: cold and heat, etc.) and plants (8:22: seedtime and harvest).

Now let us proceed to distinguish some of the aspects of the Covenantal Bond. The most obvious distinction is between the *parties* of the Covenantal Bond and the *"pact"* between them. The parties are God and his people (with the rest of Creation peripherally involved). The pact is the Covenantal Bond itself. We can focus on any one of these.

Description. The *First Polar View* of the Covenantal Bond is the View focusing on what God himself does with reference to the Covenantal Bond. The *Second Polar View* focuses on what happens to his people. The *Axial View* focuses on the relation between the parties, that is, on the Covenantal Bond itself.[28]

A second way of distinguishing is in terms of the way that the Covenantal Bond comes to Israel. There are Covenantal words (promises, historical recitals, commands; cf. Deut. 4:13), Covenantal administration ("keeping covenant"; cf. Gen. 17:9; Exod. 19:5; Deut. 7:9, 12, etc.), and Covenantal sanctions (blessings and curses; cf. Deut. 27–30). All three of these elements occur in the ancient Near Eastern Hittite treaties, which form a secular parallel to OT religious covenants (see the discussion in 3.131). All three are commonly present in biblical covenants themselves. Indeed. these three are only another way of talking about the Prophetic, Kingly, and Priestly Functions as they come to expression in the Covenantal structure.

Description. The *Locutionary, Administrative,* and *Sanctional* aspects of the Covenantal Bond are, respectively, the Covenantal words, the Covenantal administration, and the Covenantal sanctions; or, more generally, those parts of the relation between God and his people having to do with Prophetic, Kingly, and Priestly Functions of the Covenantal Bond.

Thus we have here a manner of talking about Prophetic, Kingly, and Priestly Functions as these are looked at from the standpoint of the Covenantal Bond. We can discuss the Locutionary, Administrative, and Sanctional aspects from the First Polar View, the Second Polar View, or the Axial View.

For example, taking the First Polar View: the Covenantal Bond includes God's speaking (Locutionary), God's acting to keep the Covenant (Administrative), and God's imposing sanctions (Sanctional). Taking the Second Polar View: the Covenantal Bond includes Israel's hearing God's word (Locutionary), Israel's keeping or breaking the Covenant (Administrative), and Israel's receiving the sanctions (Sanctional). Taking the Axial View: the Covenantal Bond includes God's Covenantal words (Locutionary); his mighty redemptive/judgmental acts such as the flood of Noah, the Exodus, the conquest of Joshua, the exile, etc. (Administrative); and the blessings and curses consequent to those acts, such as death in the flood, or enjoying the promised land (Sanctional). See Table 8.

Table 8

The Covenantal Bond

	Locutionary (Prophetic) aspect	Administrative (Kingly) aspect	Sanctional (Priestly) aspect
First Polar View (cf. Active)	God's speaking to Israel	God's acting for Israel	God's blessing or cursing Israel
Second Polar View (cf. Passive)	Israel hears	Israel obeys, is redeemed, etc.	Israel receives blessing or cursing
Axial View (cf. Middle)	God's words in the Covenantal Bond	redemptive acts	blessings in the Covenantal Bond

(God's speaking to Israel is included in the First Polar View of the Locutionary aspect of the Covenantal Bond. The rest of the entries are read similarly.)

3.322 *The Dominical Bond*

Now a problem arises. Not all God's relations to Creation are included under the Covenantal Bond as described above. For example, many of his words concerning snow, ice, and wind (Ps. 147:15-18) are not recorded in Scripture. And the Covenantal Bond is the pact under sanctions *revealed in Scripture*. It is limited by Scripture. This limitation is a proper limitation of "covenant" because Scripture is *the* covenant document of the people of God.[29] In ancient Near Eastern covenants, both parties knew what the terms of the covenant were. Indeed, the covenants made express provisions for preservation of the covenant words. The same is true of the Covenantal Bond (see the concern for preservation in Deut. 31: 19, 26, etc.). Though to some extent the Covenant may be forgotten by man (II Kings 22:8ff.; 17:38; Prov. 2:17; Hos. 4:6), yet it cannot be completely wiped out (Rom. 1:18ff.; 2:14-15). The Covenant words are words "shared" by God and man, to which both are bound (though, of course, God is the sovereign giver of the covenant and the arbiter of its true meaning). Some of God's words are thus not a part of the Covenantal Bond. In other words, 'Covenantal Bond' is not a broad enough category to encompass all of God's dealings with Creation.

The situation can be clarified in terms of the distinction between the Covenantal parties and the pact. God says more than the Covenantal Bond says, which in turn says more than Men hear and understand from it. Because the above description of God-Creation relations has been formulated in terms of *Covenantal* relations, it did not include everything that it might have if it had started from what *God* says and does. The perspective would have been still more anthropocentric if we had begun with God's people.

Description. The *Dominical Bond* is the totality of God's relations to himself and to Creation. Thus it includes the Covenantal Bond. A *Servient Bond* is that part of the Covenantal Bond that pertains to a given Creature.

Roughly speaking, the Covenantal Bond is that part of the Dominical Bond officially and formally shared with or revealed to Creatures in the biblical covenant. Of course, we do not and cannot know all about the Dominical Bond. We cannot know all that God knows about his relations to Creation. However, we can know some things about the Dominical Bond not explicitly recorded in Scripture. For example, God has caused the snow to fall in our lifetime, he has brought about the defeat of Hitler, etc.

I add the third part about Servient Bonds because a given person may understand more or less of the Covenantal Bond (depending on where he is in the history of redemption, and also on how much the Holy Spirit enables him to understand the Covenantal words then available). A Creature may also be affected more or less by Covenantal Administration and Sanctions.

Description. The *Bond* is the Dominical Bond, or the Covenantal Bond, or a Servient Bond.

What similarities are there between the Dominical Bond and the Covenantal Bond that would lead us to using a common term? The most obvious similarity is that the Covenantal Bond is in fact one aspect of the Dominical Bond, particularly as the Dominical Bond relates to the people of God. Moreover, the parallels are sometimes more explicit. For example, Jeremiah 31:31-37 compares the new covenant with the house of Israel to "the fixed order (ḥuqquōt) of moon and stars." What is this "fixed order"? Psalm 148:6 uses the same word as Jeremiah 31:36 to answer: "He established them [heavenly Creature] for ever and ever; he set a law (ḥoq) which cannot pass away" (cf. Ps. 104:9). Here we have to do with some of God's words respecting the pattern of events among Inorganic Creatures (probably the decree of Gen. 8:22 is in mind).

Similarly, Psalm 147 makes a comparison between God's speech to the Inorganic Kingdom (vv. 15-18) and his speech to the nation of Israel (19-20). The shifts between general providence and Israel recur throughout the psalm. Admittedly, there are *differences* between works with the Subhuman Kingdom and works with Israel, but the parallels are exploited throughout the psalter (see Ps. 148; 19:1-6, 7-14; 119:89-91, etc.).

Finally, what about Angels? Can we properly describe God's relation to them under the rubric of "Bond"? In a way, it does not much matter, since we do not know all that much about angels. However, Scripture does give us enough information to enable us to see parallels between the situation with angels and the situation of those Creatures that are definitely under some covenantal relation. The angels, like the reprobate, are under some kind of moral law (the first commandment, if none other, would be relevant to them). Violation of the law resulted in imposition of sanctions (Jude 6). The angels who did not fall away from God apparently have been confirmed in righteousness as a blessing-sanction on their obedience (see I Tim. 5:21). Angels, with other covenant Creatures, are placed under the command to praise the Lord (Ps. 148:2, cf. v. 6). They are compared to Inorganic Creatures (wind, flame) in the matter of obedience to God's word (Ps. 104:4; 103:20-22; Heb. 1:7). The very word 'angel' ("messenger") draws a parallel to the messengers of God's

Covenant (Mal. 3:1; Luke 20:9-12). Hence I include the relation of angels to God under the "Dominical Bond."

Now, as in the case of the Covenantal Bond, the Bond as a whole can be described in terms of Locutionary, Administrative, and Sanctional aspects. God speaks to someone (Locutionary), he rules by the execution of his will for or in someone (Administrative), and he opens himself in communion with blessing or cursing to someone (Sanctional). This language is language about the Prophetic, Kingly, and Priestly Functions of God—except that now the second party to whom God directs his activity is explicitly in focus (unlike in 3.1). Thus there are fuzzy boundaries between "modal" (3.1) and "structural" (3.3) analyses.

Finally, note that we can discuss the Bond from the First Polar View, the Axial View, or the Second Polar View.

Description. The *First Polar View* of the Bond focus on God, the *Axial View* of the Bond focuses on the Bond itself, and the *Second Polar View* focuses on Creation.

For an even more general, expanded use of 'Polar' and 'Axial,' see 3.332.

3.323 *The Bond and the Mediator*

Now the discussion of Bond should be brought into relation to the earlier discussion of the Mediator (2.3). Jesus Christ is not only called "mediator of the covenant" (Heb. 8:6; 9:15; 12:24), but twice is actually said to be (function as) the covenant (Isa. 42:6; 49:8). The Covenantal Bond is from first to last involved with the Mediator—we might even say that it is the manifestation of the Mediator to his people. This would be especially true if the OT theophanies were a proleptic manifestation of the Mediator (see 2.3). Covenantal Locution is the Prophetic work of Christ, Covenantal Administration is the Kingly work of Christ, and Covenantal Sanction is the Priestly work of Christ.

Jesus Christ is both God and Man. Therefore it makes sense to consider his person and work with respect to the Dominical, Covenantal, and Servient Bond.

3.3231 *Christ and the Dominical Bond: Christ is Lord*

Christ together with the Father and the Holy Spirit is Lord of Creation, commanding all things (Lam. 3:37), ruling all things (Ps. 103:19), judging all things (Ps. 33:13-15, etc.). We can consider Christ's Lordship itself from the First Polar, Axial, and Second Polar Views. Namely, we can consider Christ in relation to the Godhead (First Polar View), in relation to his mediatorial role even in creation (Axial View), and in relation to Creation (Second Polar View).

First, the First Polar View. The Father and the Son share in eternal fellowship of glory (John 1:1; 17:5). If we are so bold as to use the analogy of John 17:21 (and reason backward from God's relations to men), they share an eternal fellowship in the Holy Spirit (II Cor. 13:14). We are speaking therefore of the "ontological" Trinity. The eternal fellowship includes communication of the Spirit's words (John 1:1; and analogy on I Cor. 2:10-13), a common sovereignty as Lord (Heb. 1:8; analogy with Heb. 1:10; I Cor. 8:6; II Cor. 3:18), and the sharing in the blessing of the Holy Spirit himself (analogy with Acts 2:33).

Next, the Axial View. The Father works all things in Creation *through* the divine Son (I Cor. 8:6). Here we are speaking of the "economic" Trinity. As in the case of the First Polar View, we can distinguish Locutionary, Administrative, and Sanctional elements. The Father has through the Son decreed or commanded both the original creation of heavens and earth, that Creation should be as it is, and that all things should take place as they do (Col. 1:16, 20; John 1:3; Eph. 1:11). Through the Son he administers and executes his will in all Creation (Heb. 1:3; Col. 1:20). Through the Son he communes with men, judges, and brings home sanctions (John 14:9-24; 5:22, 27; Rev. 19:11ff.).

Lastly, the Second Polar View. Christ is Lord who decrees all things (Heb. 1:3; Lam. 3:37), rules all things (Acts 10:36; Neh. 9:6; Dan. 4:35), and is in all things (Col. 1:17; Eph. 4:10, 6; Jer. 23:24) to judge (Jer. 23:24; Rom. 2:16; Rev. 2:2, etc.). This View thus emphasizes Christ's identity with the Father. See Table 9.

Table 9

Active structure with respect to Jesus Christ

Undiffer-entiated	God	Pr speaks to Ki rules Hi blesses, appraises	Creatures
Dominical 1. First Polar View	Father	Pr eternally communicates with Ki eternally shares rule with Hi eternally blesses	Christ
2. Axial View	Father	through Christ Pr decrees Ki rules Hi appraises	all things
3. Second Polar View	Christ	Pr decrees Ki rules Hi judges	all things
Covenantal 1. First Polar View	Father	Pr decrees to send Ki sends Hi blesses with the H.S.	Christ
2. Axial View	Father	through Christ Pr gives the Scriptures Ki works miraculous works Hi gives redemption/ judgment	for his people
3. Second Polar View	Christ	Pr authors the Scripture Ki rules Hi redeems and bestows spiritual gifts	for his people
Servient 1. First Polar View	Father	Pr speaks (esp. in OT) to Ki empowers Hi gives the H.S. at baptism	Christ
2. Axial View	Father	through Christ in the flesh Pr speaks his words to Ki brings the kingdom to Hi blesses with salvation	his people
3. Second Polar View	Christ in the flesh	Pr speaks words to Ki works miracles for Hi heals, blesses	his people

(The analysis is in terms of the Dominical, Covenantal, and Servient Bond, in terms of three Views, and in terms of Prophetic (Pr), Kingly (Ki), and Priestly (Hi) work.)

3.3232 *Christ and the Covenantal Bond: Christ is Mediator and covenant*

Here we focus not on those works that Christ does as God and Lord, but those works that he does *specifically* as Mediator to redeem fallen men. Once again, let us subdivide into First Polar, Axial, and Second Polar Views, focusing on Christ's relation to the Godhead, his mediatorial role, and his relation to Creatures.

First, the First Polar View. As Mediator, what is Christ's relation to the Father? "The Father has sent his Son as the Savior of the world" (I John 4:14). He decreed to send the Son (Locutionary; Ps. 40:8), he sent the Son (John 9:4, etc), and he blessed him with the Holy Spirit without measure (John 3:34).

Second, the Axial View. Through Christ the Mediator, the Father gives the Scripture to his people, works his miraculous works, and gives redemption and judgment to Creation (Eph. 1:3; Rom. 8:21). There might seem to be some difficulty in the OT, since at that time Christ was not yet Incarnate. However, the theophanic form of OT revelations, as well as the occurrence of human mediators like Abraham (for Sodom and Gomorrah), Joseph (interpreting Pharaoh's dream), Moses, David (I Chron. 28:19; II Sam. 23:1-2, 5), prophets, and priests, are anticipations or "Preparation" for the Mediatorial appearance of Christ in the NT. Hence all the giving of OT Scripture should be understood as Mediatorially qualified—else how could sinful man hear the word of God and live (Exod. 20:19; Deut. 5:23-31)?

Lastly, the Second Polar View. Because the Mediator represents God to the people, we can straightforwardly speak of God's works in Scripture as from the Mediator. Christ is the true author of Scripture, the ruler of his people, and the bestower of every spiritual gift. In the OT, this means that we can speak in terms of redemption by the angel of the Lord (e.g., Isa. 63:9) or by the name of the Lord (Exod. 23:21; Ps. 33:21; 124:8).

3.3233 *Christ and the Servient Bond: Christ is Man and servant*

Christ is a man, and as such he is subject to God's Covenant (Gal.

4:4). The treatment of this topic is similar in structure to the treatment of 3.3231 and 3.3232. Hence a brief sketch may suffice. The First Polar View: Christ is the perfect servant who does the Father's will. The Axial View: Christ is the last Adam, the representative man, who by his sacrificial substitution has purchased salvation for a whole race of descendants (Rom. 5:12-21). The Second Polar View: Christ is the model in whose steps we follow (I Pet. 2:21-24).

We have talked of Christ and his work primarily in terms of the Active Function (3.1213). This is the normal way of talking, but we could go through the whole discussion focusing on the Middle or Passive Function. Corresponding, for example, to God speaking to, ruling, and appraising Creatures, we could speak Passively of the Creatures as listening, obeying, being appraised (receiving value). This is particularly important in the Covenantal and Servient cases, since Man's response to Covenantal commands can be either obedient or disobedient.

3.324 *The Bond and the word of God*

Now we come to the question of "law." What is the law? This is a question that no scientist can ignore.[30] But this question poses the same kind of difficulty as the question "What is there?" and "How does everything function?" It is too vague to require one and only one answer. Suppose first that we restrict ourselves to God's law. Even here Scripture uses several terms—'statute,' 'commandment,' 'judgment,' etc. In some contexts these terms can be practically synonymous with one another (Ps. 119), but in general each has its own nuances of connotation. We can ask, too, whether we are speaking of Covenantal law in the sense of something written down to which men have access, or Dominical law by which God rules all the earth. Dominical law would then be broader than (though it would include) Covenantal law. Furthermore, do we want to talk specifically about commandments (like the ten commandments), or do we want to include historical accounts (I Samuel–II Kings) as well?

The term 'law,' in other words, can be used in various senses (cf., e.g., Pauline usage of *nomos*), none of which is intrinsically "wrong."[31] The only requirements are that one make it clear to the listener what sense one is using, and that one not draw unsupportable conclusions by surreptitiously passing from one sense to another.

Hence we cannot solve all problems by a mere "definition" of law. It is a question of the characteristics and variation in God's relation to Creation. We have already considered these relations in 3.321, 3.322, and 3.323. The discussion might stop there, except that there remains a certain need to clarify certain dissimilarities between what I am saying and what other philosophers have said (see 8.217).

3.3241 *The Law of God*

Everything that the Bible calls "law," "precepts," "commandments," etc., is part of the Locutionary Bond. Almost always "law" is specifically Covenantal law, i.e., law that has been Lingually spoken to or written down for Men. Moreover, there is often a distinct connotation of obedience and command, so that the law could be seen also as Administrative and Kingly.[32] So, we can describe a technical term as follows:

Description. The *Law* is the Covenantal Locution of God as king.

Or, more narrowly,

Description. The *Law* is the Covenantal command of God.

The first of these two descriptions includes in "law" the whole Bible, while the focus is on commands—kingly rule over men. The second includes *only* law in the narrow sense of standards for human behavior. From here on the first description will be used.

3.3242 *The word of God*

The terms in the Bible for 'word' (*'imrāh, dābār, logos*) and 'statute' (*hōq, huqqāh*) have a somewhat broader range of meaning than "law." They can be used of the not-specifically-Covenantal declarations of God (Ps. 33:6; 147:15; 148:6, 8; Job 28:26; 38:33; II Pet. 3:5), and *logos* is used of the second person of the Trinity (John 1:1; I John 1:1; Rev. 19:13). Of course, it may be just an accident of usage that other words for "law" (e.g., *tôrāh, mišpāṭ, miṣwāh, piqqûdîm, nomos, entolē*) are apparently never used in describing God's words outside his *Covenantal* Locutions. Psalm 119:91 (*mišpāṭ̣ę(y)kā*) appears to be an exception, but it may very well be an allusion to the Covenantal Locutions of Genesis 8:22; 9:11, 15.

A description roughly (but of course not exactly) corresponding to this usage of 'word' might be the following.

Description. The *Word of God* is the Dominical Locution of God. Or, less technically, the *Word of God* is what God says.

Thus all Laws are the Word of God; but not all Words of God are Law.

In one sense, all of God's relations to Creation can be dealt with in terms of the Word of God. For everything that God does he does by speaking (Ps. 33:6, 11; Eph. 1:11; Heb. 1:3). On the other hand, all of God's relations could equally well be considered from a more Kingly or Administrative viewpoint. The Bible talks in terms of God's ruling as well as in terms of God's speaking, and God rules all of Creation in every detail (Ps. 103:19; Eph. 1:11; Ps. 115:3; 47:7). Thirdly, all of God's relations could be subsumed under the Sanctional perspective. Even the creation of the world is for man's *good* and the *glory* of God. Creation marks a beginning of a *communion* of God and Creation (Acts 17:28).

The comprehensiveness of any one of the three perspectives is related, I think, to the Trinitarian character of God. We have already seen that there is a correlation between the Father, Son, and Holy Spirit, and the Prophetic, Kingly, and Priestly Functions (3.131). The comprehensive character of any one of the Prophetic, Kingly, and Priestly Functions of God can then be related to the fact that Father, Son, and Holy Spirit are each fully God.

3.3243 *The "nature" of the Word of God*

Still, the description of 3.3242 may not satisfy a person. Someone may be inclined to ask: what is the "nature" of the Word of God? Is it God or a Creature or both or the "boundary" between Creator and Creature, or a "third mode of being," "neither the divine being nor . . . created"?[33] I have already partly answered this question in 2.3. The Word of God is what God says. As such, the Word has divine authority, power, perfection, truth, unchangeability, goodness, and beauty. The Word is worthy of our complete religious trust, submission, and admiration, the kind of attitudes portrayed in Psalm 119.

This is true of *any* Word of God. God need not have said everything that he has said, and indeed we need not have *heard* everything that he has said, in order for us to give trust, submission, etc., to a particular Word of God in the Bible. The reason for this is that God's Words are like the Words that God spoke from Sinai. To hear those Words is to hear God. As with a Human person, so with God, we obey God and acknowledge him as Lord by obeying his Word and acknowledging that his Word is Lord of our lives.

Next, some Words of God addressed or spoken to Man are in human language (the Bible), so that Man can understand and respond and bless God. God's Word in human language is like other pieces of human language in that it does not say everything, it uses vocabulary common to men, its language becomes archaic with the passage of time, and so forth.

But beyond this, there are varieties of Words of God. God says many different kinds of things. And if we want to become more specific, we could say that the "nature" of the Word of God (whatever "nature" may mean) is different depending on what it is that God says. Why should we expect it to be otherwise? If we think that Human language is rich in capabilities, how much more God's Words! "Thou hast multiplied, O Lord my God, thy wondrous deeds and thy thoughts toward us; none can compare with thee! Were I to proclaim and tell of them, they would be more than can be numbered" (Ps. 40:5).

But to be specific. Let us go through the "Pr" or Locutionary part of Table 9, focusing this time on the role of the Word of God rather than the role of Jesus Christ.

3.32431 *The Dominical Word of God*

The Dominical Word of God is *all* that God says. What God says can be viewed in its relation to God (First Polar View), in its role in the relations of God to Creation (Axial View), and in its relation to Creation (Second Polar View).

3.324311 *The Word of God as God (First Polar View)*

In First Polar Dominical Locution God speaks eternally his Word the Son who is God. So the "nature" of this Word of God is to be God and with God from the beginning. There are no other Words "beyond" this Word. When God speaks to his Son, he "holds nothing back" (Col. 2:9; Matt. 11:27). Everything that God is, the Word is. So there can be no talk of "going beyond" the Son of God.

3.324312 *The Word of God to Creation (Axial View)*

In Axial Dominical Locution God says, decrees, ordains, commands all the truth concerning all Creatures (Lam. 3:37-38; Eph. 1:11; Ps. 33:6, 11; 148:5, 8; 147:15, 18-20; Heb. 1:3). This includes the truth of his Covenantal obligations on Man. In the ten commandments he tells us what he approves concerning Man.

Now God's decrees and commandments are in a certain sense "relative" to Creation. They are spoken to and about Creation. Thus, from the Second Polar View, they are in a certain sense Creaturely. For example, the commandment "You shall not commit adultery" would make no sense in a Creation without sexuality. On the other hand, God's Words to Creation are also *God's* Words (First Polar View). For example, the Word has divine authority; it is always true. In particular, this means that the Creation always conforms to what God says. And this is *not* because God "looks ahead" to see what Creation will be "on its own," but because it is God's prerogative as *God* to tell the Creation what it shall be.[34] Moreover, God's prerogatives extend to Man and every part of him in Creation. "The king's heart is a stream of water in the hand of the Lord; he turns it wherever he will" (Prov. 21:1). Cyrus issued his famous decree because God said that he would (Ezra 1:1; Isa. 44:28; 45:13; Jer. 25:12; 29:10). (On this decree, see also 8.219.)

The Axial Dominical Locutions of God interlock with the First Polar Dominical Locutions, as we would expect that they should.

By that I mean that what God says to and about Creation (Axial) conforms to what he says to his Son (First Polar). What he says agrees with who he is. We could not conceive of God lying or speaking unholy things, or condoning adultery. The commandment against adultery is bound up with the fact that the human marriage relationship mirrors the relation of God to his people, and God is a jealous God (Exod. 20:3, 5). There is an analogy between the mystery of the incarnation and the mystery of God's speaking to Creation. In the case of the incarnation, the Son of God, remaining fully God (John 1:1-18), became also a man. "Remaining what he was, he became what he was not." Similarly, in the case of Scripture, God's Word, remaining fully divine, becomes Hebrew and Greek. When God speaks to sun or moon, his Word remains what it was (forever fixed in the heavens), but becomes what it was not (addressed and articulated to sun or moon). This is not a problem, because God *is* God and can do new things (see 3.32434).

3.324313 The Word of God as the "structure" of Creation (Second Polar View)

Third, take the Second Polar View of God's Dominical Locution. God says what and how and why things shall be in each aspect and part of Creation. We could go through all the subject-matter of chapters 2 and 3 discussing their relations to the Word of God. A brief summary of this will suffice here.

3.3243131 The Word of God governs and determines ontology

God declares that there shall be Angels, Men, animals, plants, and Inorganic Creatures. He declares how many they shall be, what kind, etc.

3.3243132 The Word of God governs and determines methodology

God's Word not only says *what* Creation shall be, but also "how it functions." In Dominical Locution God says that Men shall be different from animals in being the image of God and Prophets, Kings, and Priests. God says that history shall develop from Preparation to

Accomplishment to Application of redemption. He says that if anyone repents he will be saved (Ezek. 18:5ff.; 33:1ff.), that if one
keeps the commandments he will live (Lev. 18:5). Hence, in particular, the responsibility of Men to make choices that will affect their
future courses is established and ordained by God.

3.3243133 *The Word of God governs and determines axiology*

Finally, God says or declares what is good and bad in his sight
(remember Appraisive creation in Genesis 1 [3.27]; see also chapter 4). This includes his appraisal of unfallen Creation as very good
(Gen. 1:31), his appraisal of some later events as evil (Isa. 47:5;
Lam. 3:38), and his appraisal of Human actions, dispositions, and
goals as right and wrong (e.g., the ten commandments, and the final
judgment: Rom. 2:16; II Cor. 5:10). His declarations concerning
value are true and holy and just. He is saying, declaring, or ordaining
in (for example) the ten commandments *that* such and such actions,
dispositions, and goals are approved in his sight, *not* that men will
in fact *do* what is approved. Hence there is no question of a "lessen-
ing" of divine sovereignty here. The ten commandments are not in
competition with Words concerning what will in fact be, or how it
will come to be. God declares that the crucifixion will take place,
and at the same time pronounces the officials unjust (Acts 2:23).

That there is a difference between his saying, "It is good," and
saying, "Let it be," is apparent from the fact that he calls evil some
things that he does. "Does evil befall a city unless the Lord has done
it?" (Amos 3:6). At the same time, of course, even actions that he
thus calls evil turn out for his glory and the good of his chosen
(Rom. 8:28). In this respect, they may be called "good" (Ps. 119:
68). There is never anything *wrong* in what God does, because all
his actions are holy, and so he approves them all (Ps. 119:137).
Man is responsible to do what God approves, in accordance with
God's Word concerning right and wrong.

Most perversions of the biblical teaching on divine sovereignty and
human responsibility can be traced to a Reductionism that eliminates
one or more of the above three "sides" to God's Word. The elimination of the "ontological" side results in Pelagianism or Arminianism,

the elimination of the "methodological" side results in fatalism, and the elimination of the "axiological" side results in amoralism.

3.32432 *The Covenantal Word of God*

Some of the speech of Axial Dominical Locution is also Covenantal Locution, now written down for our benefit (Rom. 15:4). The Bible, then, is the Covenantal Word of God. Even here, however, one must not exclude Words spoken to the OT saints which may now be lost, or Jesus Christ himself come in the flesh. One can speak of Jesus Christ not only as the eternal Word with the Father (John 1:1), but as the Word Covenantally made known to us in the fullness of time (John 1:14). However, since the Bible is the Word of Christ, there is no need to play Christ off against the Bible. In the period following the foundation-laying of the church (Eph. 2:20), the Bible is all the Covenantal Word of God.[35] Christ is present with us because the Bible is *his* Word. He is also present, of course, in connection with his Covenantal Administration and Sanction, and in connection with the work of the Holy Spirit. These all interlock with one another.

We can now further proceed to break things down in terms similar to Table 9. For example, God says what Scripture says (First Polar View), God speaks through Scripture to his people (Axial View), and the Scripture gives life and direction to his people (Second Polar View: cf. Ps. 119).

3.32433 *The Servient Word of God*

First, the Christ of the Gospels is the servant of the Lord. As such, his preaching is a Servient Word of God. Second, Christ appointed his servants the inspired apostles and prophets who preached and wrote a Servient Word of God. Of course, what they wrote officially is also the Covenantal Word of God.

Third, even in our day preachers and teachers proclaim the Word of God. Insofar as what they say is what the Bible says, they speak a Servient Word of God which their hearers ought to receive with reverence and fear. But their words must always be checked for their conformity to the Covenantal Word of God.

Of all these servants one may say that God speaks to them (First

Polar View), God speaks through them (Axial View), and they speak authoritatively for God to Men (Second Polar View).

3.32434 *What the Word of God is "really"*

Some will probably feel that I have not yet answered the really burning question about whether the Word is a "third mode of being" or a "boundary."[36] I think that I have answered the question, more by treating it as a confused and inapt question than by giving it a "yes" or "no" answer. However, if one must have an answer, the answer is that I would not call the Word either a "third mode of being" or a "boundary" between God and Creation. Both of these expressions sound too much like a denial of the divinity of God's Word, and of the Creatureliness as well of such Words as are spoken to Creation. Moreover, neither "answer" ("third mode" or "boundary") says much positively or clearly.

Then the question reasserts itself, "What *is* the nature of the Word of God?" But now I challenge the propriety of the question on three grounds.[37] (1) What does the questioner want to know *beyond* what has been said in 3.3243? What does he *need to know in order to live* Christianly or to talk meaningfully about the "word of God"? I doubt whether he can give a clear sense to the question. (2) The questioner may not understand one of the "ground-rules" for the Christian faith, namely that God's answers *ought* to satisfy inquirers. If they do not satisfy, it is the inquirer's fault. We do not need, nor are we to seek, answers somehow deeper and more "ultimate" than the Bible's answers. Of course, my answers can satisfy in this way only insofar as they are God's answers.

(3) The question is a confusing one even in the case of human language. Suppose Abe is talking to a friend Bill on the telephone, and Charlie asks Abe, "What are those words coming over the telephone?" It's a rather odd question to begin with.

Abe says, "I'm listening to Bill tell me about the game last night."

C: What is the nature of those words that you are listening to?"

A: "They're Bill's words."

C: "Yes, but what are they like?"

A: "They're like the way Bill speaks, of course."

C: "What is their *nature?*"

A: "He's talking in English, if that's what you mean. With college-educated grammar, and Mid-Western accent. He's describing the game, as I said. Bill's quite a football fan."

C: "Aren't you *really* just listening to that receiver and the air-waves rather than to Bill?"

A: "Look, I can explain to you about electric currents, micro-phones, sound waves, and the like, but I don't think that that's your problem. I *am* listening to Bill. Now what's the problem?"

C: "I want to know what language *really* is."

A: "As a rough and ready answer, I'd say it's what we communicate with."

C: "That's what we use it for, but what *is* it?"

What can Charlie want to know that he does not already know? We could put Charlie through a sequence of courses in linguistics (as, analogously, I have tried to do in a sketchy way with the Word of God in 3.3243). But after it was all over he could say, "That is how language is structured and related to other things, but what *is* it?" One must then seriously ask whether Charlie is striving for a God-like knowledge of language.

3.33 *The structure of Creation*

So far, we have considered God's relations to himself (briefly) and God's relations to Creation. Now let us consider in greater detail relations of Creatures to one another. This does not mean that God is out of the picture completely, but only that we shift into the foreground relations that have until now been in the background.

Now how shall we deal with relations among Creatures? It should not be imagined that there is only *one* right way of dealing with such structure. The approach of Proverbs is different from the approach of the Books of Moses. Because God is so wonderful, and his decrees respecting Creation so rich (3.3243), we ought primarily to admire the richness of God's Creation, rather than to imagine that we can completely sort things out. Nevertheless, some classification is useful if only to reawaken us to the richness, and to guard against the Reductionisms that are as common here as they are in the area of modality (3.133).

This also means that not any classification will do, because it is easy to follow the lead of secularists in wiping out differences between different types of relations.

3.331 *The church*

The most obvious scriptural starting point is the federal headship of Adam and of Christ, described in Roman 5:12-21, and presupposed in many more passages of the NT. This idea is not unique to the NT, since Moses, David, and others represent Israel before God, thereby prefiguring the representative character of the work of Christ as man.

Now Christ is the head of his people, so that what happens to him also involves them. "As in Adam all [who are represented in Adam] die, so also in Christ shall all [who are represented in Christ] be made alive" (I Cor. 15:22). When he dies, he dies *for* his sheep (John 10:15), so that they also have died (II Cor. 5:14). His righteousness is accounted to them (II Cor. 5:21). The relation of Christ to his own is also explained with the figure of the body: the church is the body of Christ, and each of us a member of it (I Cor. 12:12ff.). Or again, Christ is the vine, and his people the branches (John 15). The richness and frequency of the imagery (think of Paul's "in Christ"!) shows how rich a relationship we are dealing with. We are considering primarily what in terms of Table 9 would be called the Second Polar View of the Covenantal Bond and the Second Polar View of the Servient Bond. Christ as Servant of the Covenantal Bond is identified with his people.

Now, if our main interest were ecclesiology, we might proceed to analyze this "federal" structure more closely in terms of its Locutionary, Administrative, and Sanctional aspects, or in terms of a focus on the Head, on the relationship itself, and on the members (these three would be analogous to the earlier First Polar, Axial, and Second Polar Views). We might also trace the development of the people of God through the Periods delineated in 3.2.

But instead, let us use this federal structure as a starting point for dealing with structure in general. Since the church is a new humanity, a restoration (and more) of what was lost in Adam and the "old" humanity that he represented, we might naturally expect that some of the structure of church relationships would be similar to what we can still find (though in sinful form) among the Cosmic Human Kingdom.

3.332 *The Bond and Creation*

Let us now trace out some of the ways that this is so. In the first place, as in the case of the First Polar, Axial, and Second Polar Views of the Bond (3.322), it is true for any structure that we can consider it from the standpoint of one or more of the parties, or from the standpoint of the structure itself. This is obviously a matter of degree, just as it was for the First Polar, Axial, and Second Polar Views of the Bond.

> Description. *Polar Views* of a structure are views with a focus on one or more of the parties; *Axial Views* are views focusing on the relations between or among parties.

As an example, take the comparatively simple case of the relation among mountains in a mountain chain. We can take a Polar View that focuses on one mountain and asks how the others help to explain the structure of this one mountain, how the other mountains influence the weather patterns and erosion on this one, etc. Or we could take an Axial View by focusing on what the mountains have in common, how they are situated with respect to one another, etc.

A second kind of division of perspective relates more directly to the Dominical, Covenantal, and Servient Bond. Namely, we can ask about the view of a structure in terms of all its relationships, or about Scripture's view, or about a Servant's view.

> Description. The *Dominical, Covenantal,* and *Servient* Views of a structure are views from the standpoint of the Dominical Bond, the Covenantal Bond, and a Servient Bond respectively.

Thus, for example, in the Dominical View we include that God declares what the relation among the mountains shall be, that in his providence he maintains and alters the structure, that he is pleased to use the relations for his purposes, and so on. In the Covenantal View, we look at what Scripture says about mountains and about these mountains in particular. In the particular case of mountains, not all that much is said—except if one happened to pick Mt. Sinai, Mt. Zion, or the like. In the Servient View, we look at how a particular man or men understand and appreciate the relation among the mountains.

3.333 *Application of modality to structurality*

But all this may seem to be not completely adequate. It amounts to a kind of describing of relations in terms of relations of relations. and these in terms of relations of relations of relations, and so on. The problem is that we have been using a Field View (which, of course, is the way one "gets hold of" structure; cf. 3.123). But for *classifying* structures, a Particle View is needed. The classification terminology of 3.1 therefore becomes uesful.

First, recall the Particle, Wave, and Field Views developed in 3.123. Any structure can be viewed in these three ways. The church, for example, we can talk about in terms of being a unified whole with boundaries to its membership (Particle View; though perhaps this has less appeal when there are so many denominations!). Or we can look at the church in terms of the *process* of the application of redemption which God is working in her (Wave View). Or we can view the church in terms of the relations of its members to one another and to their Lord and to the world (Field View).

In the case of many structures, one of the three Views has a kind of prominence (though this is obviously a matter of degree). We meet with unified wholes in contrast to other wholes (Particle prominence); with processes (Wave prominence); with relations (Field prominence).

Description. A *Particulate Unit* or *Thing* is a structure that we regard normally as a unified whole enduring as more or less the same over a time span. An *Undulatory Unit* or *Transaction* is a structure that we regard normally in terms of process, as a unified whole of events. A *Relational Unit* or *Relationship* is a structure that we regard normally in terms of relations among things, enduring more or less through time.
Description. A *Unit* is a Particulate, Undulatory, or Relational Unit.[38]

Remember that these are vague descriptions. The descriptions may sound puzzling, but some examples may help to say what they are talking about. First, take examples dealing with the Economic Function. A factory is a Particulate Unit, a sale is a Transaction, and the market is a Relationship. A stock exchange could be said to be a Relationship (viewed in terms of the buyers and sellers that make it work) or (as I prefer) a Thing (viewed in terms of its unity based on written charters, contrast with other exchanges, etc.).

Second, take examples from a mammalian body. A molecule is a Thing, a nerve impulse is a Transaction, and the coordination of limbs is a Relationship.

Next, we can distinguish Units in part by what Function or Functions or Mode their most prominent characteristics belong to.

> Description. A Unit is *Weighted in* X or has X *Weight* when the X Function or Mode stands in prominence in the Unit's characteristics.[39]

For example, a factory, a sale, a market, a stock exchange all have Economic Weight. A molecule has Physical Weight. The coordination of limbs has Behavioral Weight. A number has Quantitative Weight. Clearly these are easy cases. In more complicated cases a Unit might have several Weights, or no easily discernible Weight, or people might plausibly disagree over what its Weights are ('Weight' is a vague term).

The purpose of the above distinctions has been mainly to prepare the way for a classification of human relationships. So we specify our narrower concern as follows:

> Description. A *Societal* Unit is a Unit including Men in its internal substructure.
> Description. An *Institution* is a Particulate Societal Unit.[40]

Institutions can be further differentiated in terms of the way in which the Bible requires or does not require us to participate in them. To anticipate a later distinction (chapter 4), it is a question of whether the requirement of participation is primarily of a normative character (universally binding), of a situational character (required only for those in a particular situation and with a particular calling), or of an existential character (in itself, a matter of personal preference, though naturally within the limits set by normative and situational considerations).

Description. *Obligatory*[41] Institutions are those mentioned explicitly in Scripture, which are such that, if a person belongs to the Institution in question, he ordinarily ought not to withdraw his participation except on dissolution of the Institution.

Obligatory Institutions are the state, the family, marriage, and the church. The qualifying phrases are added because of various factors: (a) not all people are married; (b) marriage is effectively dissolved by death or adultery of one partner;[42] (c) a person's family may die off; (d) the state may effectively dissolve into anarchy, or it may "dissolve" as far as a given individual is concerned when he leaves its geographical bounds (but even in that case he will generally speaking have entered another state); (e) a church may apostasize.

Description. *Strategic* Institutions are those which, in many situations, people with particular callings are virtually obliged to join in order to fulfill those callings.

Under Strategic Institutions come labor unions, business enterprises, and schools.

Description. *Voluntary* Institutions are those in which membership is normally determined by personal considerations, not tightly bound up with a man's major calling.

Obviously considerable overlap and fluidity among these three is possible. Typical Voluntary Institutions are clubs (though a country club can become virtually a business necessity) and charitable organizations (though for the salaried employee of a charitable organization it is a Strategic Institution).

Institutions can also be distinguished in terms of degree of specialization. One way of doing this is in terms of the Functions.

Description. A Unit with rather clearly discernible Weight in one unique Function of the nine Functions Dogmatic, Presbyterial, Diaconal, Lingual, Juridical, Economic, Cognitional, Technical, Aesthetic is called *Differentiated.* A Unit not discernibly Weighted in only one of these, but Weighted, say, in the Sabbatical, Social, Laboratorial, Prophetic, Basilic, or Hieratic

Table 10

A classification of some Societal Units

Unit	classification	Weighted in
state	Obligatory Institution	Juridical (Differentiated)
church	Obligatory Institution	Sabbatical (Semidifferentiated)
family	Obligatory Institution	Social (Semidifferentiated)
marriage	Obligatory Institution	Social (Semidifferentiated)
university	Strategic Institution	Cognitional (Differentiated)
school	Strategic Institution	Cognitional (Differentiated); somewhat Technical
business enterprises	Strategic Institution	Technical (Differentiated); somewhat Economic
political party	Voluntary Institution (borders on Strategic)	Lingual (Differentiated); somewhat Juridical
orchestra	Voluntary Institution (Strategic for professionals)	Aesthetic (Differentiated)
army	Obligatory-Strategic Institution	Economic/Juridical (Diff.)
labor union	Strategic Institution	Economic/Lingual (Diff.)
tribe	Obligatory Institution (as a form of the state)	Personal (Undiff.)
buyer-seller	Undulatory Societal Unit	Economic (Diff.)
speaker-listener	Societal Transaction	Lingual (Diff.)
host-guest	Societal Transaction	Economic (Diff.)
attacker-defender	Societal Transaction	Economic (Diff.)

Table 10 *(continued)*

Unit	Classification	Weighted in
teacher-student	Societal Transaction (borders on Relationship)	Cognitional/Juridical (Diff.)
performer-audience	Societal Transaction	Aesthetic (Diff.)
preacher-audience	Societal Transaction	Dogmatical (Diff.)
worshipers	Societal Relationship	Sabbatical (Semidiff.)
friendship	Societal Relationship	Social (Semidiff.)
class (at school)	Societal Relationship	Cognitional (Diff.)
gossip grapevine	Societal Relationship	Lingual (Diff.)
scientific community	Societal Relationship	Prophetic (Semidiff.)

Functions is called *Semidifferentiated*. A Unit Weighted simply in the Personal Mode or less is called *Undifferentiated*.

Table 10 gives a sample of how various Societal Units might be classified using the vocabulary that we have developed. However, because of the interlocking of Functions, one must guard against any tendency to think of these classifications as "rigid." Moreover, the classifications do not say everything. They do not indicate, for example, the close relationship (a Societal Relationship) between labor unions and business enterprises, or between political parties and the state.

The present-day confusion makes necessary a special remark about the church.[43] By 'church' I mean the people of God, especially in the form that they take in the Application Period. They are the Messianic assembly, the body of Christ, the fellowship of the Holy Spirit.[44] As such, their center, their focus, their rallying point is Sabbatical worship, Prophetic, Kingly, and Priestly. However, church activity should not be *limited* to worship, in the narrow sense, any more than family activity is limited to Socially Weighted acts. Everything that a believer does he should do as a member of the body of Christ, as a churchman. Everything belongs to him—the world, life and death, the present and the future—and he belongs to Christ (I Cor. 3:22-23). For the church to cease to appreciate its interlocking with the Social and the Laboratorial, or for it to become over-Differentiated, is to begin to fail in its calling.

3.34 *Interlocking of ontology, modality, temporality, and structurality*

It is time to look more directly at something that has been going on "under the surface" all along. The sections 2, 3.1, 3.2, and 3.3 are mutually interconnected, and the categories introduced in each of these sections can help us to understand better the richness of variety contained under any one of the sections. Let us see how the categories in each of these sections apply to the other sections.

3.341 *Ontology applied*

Ontological categories of chapter 2 apply to modality (3.1), temporality (3.2), and structurality (3.3).

3.3411 *Ontology applied to modality*

The distinctions among God, Men, Angels, etc., were discussed under "ontology." We have already applied ontology to modality in observing that the Prophetic, Kingly, Priestly, Social, Laboratorial, etc., Functions of God are in some ways different from the same Functions of Men, or of Angels. When God speaks (Lingual), his Word has divine authority, power, holiness, creativity, etc. We could also go through the Subhuman Kingdom asking what connection the Functions have with each Kingdom within it. Lingually, various animals are spoken to and about, Cognitionally God and Men know various things about various Subhuman Creatures, and so on.

3.3412 *Ontology applied to temporality*

In 3.2 we focused on the discussion of time from the standpoint of God's purposes of redemption, especially in terms of the Covenantal Bond. However, we could discuss in turn time for God (his eternity) time for men (their prehistory, birth, life, death, and judgment), time for animals (their lives), for plants, etc. The Generational, Developmental, and Culminational Views could be applied to each of these.

3.3413 *Ontology applied to structurality*

In the discussion of the Bond we have already discussed how the Bond looks from the First Polar, Axial, and Second Polar Views. But each view includes the whole relationship, looked at with a certain emphasis and interest. That is somewhat different from asking, what does God do, what does Christ do, what does this or that man do. These questions could also occupy our time. But let us move on.

3.342 *Modality applied*

In examining ontology, temporality, and structurality, one can focus on one or more Modes or Functions.

3.3421 *Modality applied to ontology*

In a discussion of the Creator or of Creatures, Modes and Functions can be used. For example, we can focus on one or another or several Functions of God. We can divide Men into classes in terms of how gifted or productive they are in various Functions.

Animals and plants can be classified in terms of various Functions. Are they beautiful (Aesthetic)? Do we know much about them (Cognitional)? Are they valuable to us (Economic)?

3.3422 *Modality applied to temporality*

We have already applied modal categories to temporality to a certain extent with the introduction of the terms Vocative, Dynamic, and Appraisive.

3.3423 *Modality applied to structurality*

We have applied modality to structurality with the terms Locutionary, Administrative, and Sanctional.

3.343 *Temporality applied*

3.3431 *Temporality applied to ontology*

We can discuss the state of the various Kingdoms at various points within history. We could begin with those parts of Genesis 1 that describe the creation of the Kingdom in question. Then we could ask what was the relation of each Kingdom to Adam before the fall. What were the effects of the curse on the Kingdom. How was, for example, the Animal Kingdom related to Israel in terms of clean and unclean? What will it mean for there to be a new earth through the work of Christ?

3.3432 *Temporality applied to modality*

Next, we can look at the variation and development of the Modes and Functions in the course of the history of redemption. We could thus do a kind of history of each Function. We would obtain "historical Physics," "historical Biology," "historical Behaviorology," "historical Logic," "historical Technology," and so on.

To prepare for questions of philosophy of science, we will focus on a historical view of the Prophetic Function, and especially the Dogmatic Function. As we have already seen, this begins with the eternal speaking of the Father to the Son, before the foundation of the world. Then God speaks in creating the world (Ps. 33:6; cf. Vocative creation in 3.27). His first recorded Covenantal act toward man is also in speaking (Gen. 1:28-30).

As God has named the Creatures in Cognitional Functioning, so man follows and imitates God by naming the animals (Gen. 2:19-20). The fulfillment of the Adamic mandate (Gen. 1:28-30) obviously involves the continual exercise of Cognitional and Lingual Functions. Adam as the federal head of the race is the leader and representative in such exercise.

Next, the fall brings a corruption of man's Prophetic Function, as Adam and Eve try to "pass the buck" (Gen. 3:10-13). Yet God's own Lingual promise is the beginning of restoration, not only of God's rights but of man's welfare (3:15, 20-21).

And so we could go on. Let us focus on a few outstanding Prophetic figures in the rest of Scripture. Chief among these are men who hold the extraordinary Dogmatic or Prophetic office, that is, men who speak inspired words from God. We think of Noah, Joseph, Moses, the unknown authors of Joshua–II Chronicles, and Elijah, Isaiah, Jeremiah. All their Dogmatic works find their fulfillment in Christ. They are like an arrow pointing forward to the final speaking of God to men through Christ (Deut. 18:18; Heb. 1:1-4). We could also mention cases where man intercedes before God on behalf of others—a combination of Priestly and Prophetic roles. Abraham (Gen. 18:16ff.), Moses (Exod. 32:30), and Ezra (Neh. 9:6ff.), in their intercessory roles, again point forward to Christ (Heb. 7:25; John 17).

At the climax of the Developmental Preparation Period appears Solomon, who does still another thing. He returns to the kind of Laboratorial, Cognitional naming that reminds us of Adam (I Kings 4:29-34). But if Solomon had great wisdom, Christ has even greater wisdom (Col. 2:3). Thus Christ fulfills this aspect also of Solomon's kingdom work.

The point of this all is that our modern Prophetic Functioning, whether scientific or otherwise, must, in this Developmental Application Period, be an application and appreciation of, an entering into, Christ's wisdom and Prophetic work. This is so because only in Christ can we learn again how to communicate uprightly (Ps. 120:2ff.; 12; James 3). The implications of this can be spelled out in more detail only after a fuller consideration of what science is.

As an additional example of applying temporal development to modality, let us consider all the Functions and Modes together. How is language concerning the different Modes used in describing the redemption of God's people? In the Gospel of John redemption is closely bound up with a number of key words: grace, love, truth, life, etc. These key terms are associated with a number of different Modes and Functions (see Table 11A). The use of several terms helps the reader to appreciate the comprehensive character of God's redemption in Christ. We can make a similar classification of Pauline terms and those systematic-theological terms that have been derived primarily from Paul. See Table 11B.

Table 11

A. Terms in Johannine writings, classified by Modes

	Prophetic	Kingly	Priestly
Social love	word, witness	advocate	grace, love
Laboratorial	truth	peace	glory
Behavioral	joy(?)		
Biotic	life, resurrection		
Physical	light		

B. Terms in Paul and in systematic theology, classified by Modes

	Prophetic	Kingly	Priestly
Social sonship, adoption		righteousness justification (*dikaioō*)	redemption (*agorazō*)
Laboratorial reformation(?)	enlightenment	peace, "pacification," reconciliation (*katallagē*)	holy, glorification, sanctification (*hagios, hagiazo, doxazō*)

Behavioral	
Biotic	strengthening
	life, vivification (*zōopoieō*)
Physical	re-creation (*kainē ktisis*)

3.3433 *Temporality applied to structurality*

Under this heading comes discussion on the historical ("temporal") development of the Covenant, including the variety of covenants in Scripture: Adamic, Noachic, Abrahamic, Mosaic, etc.

Of particular interest for the philosophy of science is the way in which the Adamic mandate of Genesis 1:28-30 is taken over and developed in later covenants. This mandate to Adam involved a command to engage in scientific and technological activity. The original mandate can be considered with respect to Man-God relations, with respect to Man-Man relations, and with respect to Man-Subhuman Creature relations (these correspond to the Sabbatical, Social, and Laboratorial creation ordinances).

First, the whole of the command in Genesis 1:28-30 is to be fulfilled as a service to God (Man-God relation). Second, the commands to "be fruitful and multiply" have obvious reference to the multiplication of the human race, and hence Man-Man relations. Third, the commands to "fill the earth and subdue it, and have dominion . . ." are more closely related to the Laboratorial ordinance and Man's relations to the Subhuman Kingdom. The two sets of commands (Social and Laboratorial) are interrelated, since Man cannot fill the earth (Laboratorial) without numerical multiplication of the race (Social); conversely, he cannot be fruitful and multiply without receiving from the Subhuman Kingdom the means to sustain the multiplied human life.

After the fall, the original mandate is not simply abolished or negated, but transformed. Man is no longer in a position to fulfill the original mandate, because he is under the power of sin. But the fact of sin does not completely destroy God's purposes. God's promise of dealing with sin comes in a form analogous to the earlier mandate concerning the Social ordinance. God will give a "seed" to the woman (3:15). With regard to the ordinance of labor, a curse hangs over the ground (3:17), yet there is at least an implicit promise that it will yield food as long as man lives (3:18-19).

The mandate of Genesis 1:28-30 is renewed in Genesis 9:1ff., and already man's dominion over the Animal Kingdom is in part restored (9:2).

In Abraham the promise given in Genesis 3:15-19 becomes more specified and developed. God promises to Abraham land (12.1) and seed (12:2). These two elements correspond to the ordinances of family and labor that we have detected in Genesis 1-2. God provisionally fulfills the promise to Abraham in Joshua's time (Josh. 24:3, 13), and later under Solomon (I Kings 4:20-21). In turn, both Joshua and Solomon are types of Christ, who is the seed of

Abraham (Gal. 3:16) and inherits the whole earth (Rom. 8:17; Ps. 2:8).

The manner of Christ's fulfillment of Genesis 1:28-30 can be traced in more detail. He fills, not the earth alone, but all things (Heaven and the Cosmos) (Eph. 1:23; 4:10). He has subdued all things under his feet (Eph. 1:22).

The task of "being fruitful and multiplying" is fulfilled both in the richness of grace that comes from Christ (I Pet. 1:2), and in the multiplication of numbers of his people, the "fruit of the travail of his soul" (Isa. 53:11; John 12:24; I Cor. 15:23; Rev. 7:9). Hence the present work of "multiplying" (making disciples of all nations) and "subduing" (teaching them to observe all that I have commanded you) must be regarded as a participation in and an outflow of the finished work of Christ. In particular, the work of science can no longer take its start directly from the Adamic mandate, as it might have apart from the fall. Now it must find its roots in the mandate of the last Adam, Matthew 28:18-20. Science should be animated by Christ's fulfillment of Genesis 1:28-30

Nevertheless, this does not imply that the original mandate of Genesis 1:28-30 now has no relevance for the Christian. Christ appeared to take away sins (I John 3:5). Now that the power of sin has been broken in the believer's life (Rom. 6:7), he is for the first time able to fulfill the mandate of Genesis 1:28-30 by serving God's glory in labor, in family life, and in Sabbatical worship. He does this by following in the steps of Christ (I Pet. 2:21ff.).

Thus, expressing the task of the Christian with reference either to Genesis 1:28-30 or to Matthew 28:18-20 involves only an Emphasizing Reductionism. We must guard against an Exclusive Reductionism that would either deny that the sinful disability of man can be remedied only through the gospel (denying Matt. 28:18-20), or separate life into a "sacred" and a "secular" realm, only one of which is the sphere for service to God (denying Gen. 1:28-30).

3.344 *Structurality applied*

A few brief remarks will suffice for indicating what topics might be treated here, if they were relevant.

3.3441 *Structurality applied to ontology*

Under this heading we could discuss the relationships among the various Kingdoms.

3.3442 *Structurality applied to modality*

The Active, Middle, Passive Functions are one example of this application. They view an activity from Polar (Active, Passive) and Axial (Middle) Views.

3.3443 *Structurality applied to temporality*

Here we might ask about the relations between different Periods. Talking about promise and fulfillment already presupposes a structure of relations between the earlier and later Periods.

3.35 *The relation among fundamental categories*

Where, now, did the distinctions ontology/methodology/axiology, or modality/temporality/structurality, or Particle/Wave/Field come from? What is the relation among these categories?

This is a difficult question. I can do little better than say that they seemed to me useful distinctions to make. Somehow we have to avoid having to speak about everything at the same time, and the types of division that I have suggested are one of concentrating on one thing at a time. I do not claim that mine is the only good way of dividing up topics. I do not want to claim very much.

On the other hand, I must admit having some deeper motives for choosing categories like these. One is that I wanted to adopt some categories which, so to speak, have built into them a denial that I was making sharp distinctions where only rough ones are possible. Interlocking, overlapping categories are what I wanted. Second, I wanted to avoid at all costs the impression of dialecticism that is so fond of dual categories: nature/grace, revelation/reason, matter/form. Unbelievers frequently use a kind of dialectic that appeals to first one side, then another, of a tension or paradox. They appeal to whichever side serves their purposes, thus justifying whatever they want, and at the same time retaining the appearance of truth. They move in "dialectical" fashion from one pole to another of a paradox. But dialectic has less surface plausibility if a third element can be introduced. Of course, "third categories" are no guarantee that we will avoid error. Hegel's famous triad of thesis, antithesis, and synthesis still retains a "dialectic."

But third categories *can* be used profitably. Thus, instead of talking of nature/grace, talk of Preparation, Accomplishment, and Application of redemption, or more specifically of creation, probation, and consummation. In place of reason/revelation, talk of Dominical, Covenantal, and Servient Locution. In place of matter/form, talk of ontology, methodology, and axiology.

For all this, we still have no absolute *guarantee* that this book is not contaminated with "non-Christian categories." But what is a "non-Christian category"? The discussion of Reductionism in 3.133 suggests that the problem of non-Christian philosophy and non-Christian science is not in its vocabulary *by itself,* but in (a) false and misleading statements and (b) ambiguous terminology that can be *used* to achieve Slippery Reductionism. Even terminology that is ambiguous *need* not be so used. Hence, if this book is mistaken, it is not *merely* because of terminology.

But let me say this. If there is something fundamentally wrong with what I have called Particle, Wave, and Field Views, then what I have written is in serious trouble. For these Views are in the background of what I have written even when they are not in the foreground.

For example, there is a vague relation between Particle/Wave/ Field Views and chapters 2, 3, 4 on ontology, methodology, and axiology. In chapter 2 I have looked at things from a Particle View: what is there? In chapter 3, from a Wave View: what is the dynamic of things, how do they vary, how do they function? In chapter 4, from a Field View: how are things related to one another? In particular, is an Item "good for" something, and is the Item approved by God? Similarly, the division within chapter 3 into "modality," temporality," and "structurality" is a kind of Particle/Wave/Field subdivision within methodology. Within chapter 4, the division into Normative, Existential, and Situational Perspectives is again an application of Particle/Wave/Field Views. The Normative Perspective views an Item as a Particle, in terms of its distinctiveness, its contrasting with or comparing to other Items; the Existential Perspective views the Item in terms of where it came from and is going to, in terms of its dynamic; the Situational Perspective views the Item in terms of what it contributes to and derives from the situation, that is, in terms of its relations.

It seems to me that some kind of case can be made for the claim that Particle/Wave/Field Views are correlated with the Prophetic/Kingly/Priestly Functions, respectively (though each obviously interlocks with all the rest). If this cannot be seen directly, it can perhaps be seen by observing the correlation between Particle/Wave/Field and ontology/methodology/axiology (mentioned above), and then also between ontology/methodology/axiology and Prophetic/Kingly/Priestly. Communication characterizes an Item (ontology), Kingly power is in its functioning (methodology), and its value (axiology) is clearly related to the Priestly Function.

We have already seen that Prophetic/Kingly/Priestly Functions are mysteriously related to the Father, Son, and Holy Spirit. Does this mean that the other triples of categories that I am using are *also* related to the Trinity? Could it be that the inability to achieve precision in or with the categories is related to our inability to talk with precision about the Trinity? I do not wish to press these questions beyond what Scripture can tell us, so I cannot answer them with assurance.

NOTES TO CHAPTER 3

1. Presumably the term 'rule' in Gen. 1:18 is a somewhat metaphorical expression, based on analogy with the more "literal" rule by God and men. The possibilities for metaphor increase the vagueness of the boundaries of Modes and (as we shall see later) Functions. A word can be used *more or less* metaphorically.

2. I say this over against the treatment of "modes" or "aspects" by Herman Dooyeweerd and Hendrik G. Stoker. Both of these men at times give the impression that, if only we could think self-consciously enough and grasp clearly enough what the "meaning-kernel" of each mode is, we would be able to decide, of any particular characteristic, which mode it is related to, or whether it marks an "anticipatory" or "retrocipatory" "analogy" with another

"mode." See Dooyeweerd, *A New Critique of Theoretical Thought* (Philadelphia: Presbyterian and Reformed, 1969), I, p. 18. Stoker is somewhat more cautious than Dooyeweerd in *Beginsels en metodes in die wetenskap* (Potchefstroom: Pro Rege-Pers, 1961), p. 166. But cf. *ibid.*, p. 44, where he says that the "meaning-kernels" are self-evident principles.

If this is what they are saying, I set over against it a different way of proceeding in which (a) there is no precise "meaning-kernel," but rather a blur with a (roughly fixed) center, (b) characteristics themselves (e.g., "reproduces," "is colored," "rules") are not precisely defined; (c) the situation is not always a clear either-or situation in which everything must fall yes-or-no-fashion into exactly one modal "box" (see Appendix 3). Readers can judge for themselves how similar my description of Modes is to Dooyeweerd's or Stoker's modes. Note that I am not using the term 'mode' in exactly the same way that they are. It is not a question of there being one "right" way to use the word 'mode.' They are free to use the word 'mode' in another way.

I do, however, view with suspicion anything with the appearance of an "infinite precision" claim. Dooyeweerd's naïve/theoretical distinction, if it is indeed a "sharp" distinction, is perhaps the cardinal example of an infinite precision claim in cosmonomic philosophy. See my remarks in Appendix 2.

3. I assume that some form of sabbath observance took place before the fall. See John Murray, *Principles of Conduct; Aspects of Biblical Ethics* (Grand Rapids: Eerdmans, 1957), pp. 30-35.

4. See "Creation Ordinances," in *ibid.*, pp. 27-44.

5. These descriptions are paraphrases of the descriptions in Kenneth L. Pike, "Foundations of Tagmemics—Postulates—Set I" (unpublished; 1971), pp. 13-14. See also Pike, "Language as Particle, Wave, and Field," in *The Texas Quarterly* 2 (Summer, 1959), pp. 37-54; *idem, Language in Relation to a Unified Theory of the Structure of Human Behavior*, 2nd revised ed. (The Hague-Paris: Mouton, 1967), pp. 510-513; *idem, Linguistic Concepts* (unpublished; 1968).

6. In my opinion, this makes it difficult to see what meaning there can be in Dooyeweerd's claim that his aspects are linearly ordered (*New Critique*, II, pp. 49-54). Hendrik Stoker introduces a distinction cutting across Dooyeweerd linearity, but he still does not abandon the linearity (*Beginsels*, pp. 164-167).

7. The literature on the relation between biblical covenants and ancient Near Eastern patterns in now extensive. See especially Dennis J. McCarthy, *Treaty and Covenant* (Rome: Pontifical Biblical Institute, 1963); George E. Mendenhall, *Law and Covenant in Israel and the Ancient Near East* (Pittsburgh: Biblical Colloquium, 1955); Meredith G. Kline, *Treaty of the Great King; the Covenant Structure of Deuteronomy: Studies and Commentary* (Grand Rapids: Eerdmans, 1963; *idem, The Structure of Biblical Authority* (Grand Rapids: Eerdmans, 1972), pp. 27-44; K. A. Kitchen, *Ancient Orient and Old Testament* (Chicago: Inter-Varsity, 1966), pp. 90-102.

8. Mendenhall, *Law*, pp. 31-34; Kitchen, *Ancient Orient*, pp. 92-94.

9. Abraham Kuyper, *The Work of the Holy Spirit* (Grand Rapids: Eerdmans, 1941), p. 19. The term 'bring forth' may call to mind the planning that

we associate more with the Father, or the "bringing forth" of creation by the word of God (Ps. 33:6), or the prophetic pronouncement that typically precedes accomplishment ("arrangement"). Hence the association of Prophetic Function with the Father.

10. "It is the same God from whom, through whom, and by whom are all things, who is at once the Father who provides, the Son who accomplishes, and the Spirit who applies, redemption."—Benjamin B. Warfield, "God," *Selected Shorter Writings of Benjamin B. Warfield*—I, ed. John E. Meeter (Nutley, N. J.: Presbyterian and Reformed, 1970), pp. 69-70.

11. This is so even *if* the term 'sovereignty' were appropriate for the "spheres" of the cosmonomic philosophy of Dooyeweerd, Vollenhoven, and Stoker. I appreciate some of the positive points that were made by Abraham Kuyper's "sphere sovereignty." With this tenet he opposed the totalitarian claims of the state and of the Roman Catholic Church. Even in Kuyper, however, "sphere sovereignty" tended to restrict the authority of church officers (pastor, elder, and deacon) to the "institutional" church, in distinction from the organ body of Christ. I am convinced that this is a poor way to draw the line, and in our day "sphere sovereignty" is used to deduce unbiblical conclusions that Kuyper did not foresee. Cf. John M. Frame, *The Amsterdam Philosophy: A Preliminary Critique* (Phillipsburg, N. J.: Harmony Press, *c.* 1972), pp. 46-49.

12. To this it is usually objected that God does not cause sin—and James 1:13 is cited. The word 'cause' is admittedly blunt and open to misunderstanding. But Scripture uses the language of cause: Isa. 63:17; I Kings 12:15; Josh. 11:20; II Thess. 2:11-12; Heb. 1:3. God does not *approve* sin. Nor does he tempt men. Rather, he causes them to be tempted or not (Matt. 6:13; 4:1). For a further discussion of the compatibility of these actions, see 3.323, 3.324.

13. This form of criticism is more "transcendent"; see Herman Dooyeweerd, "Transcendent Critique of Theoretical Thought," *Jerusalem and Athens,* ed. E. R. Geehan (Philadelphia: Presbyterian and Reformed, 1971), pp. 74-77.

14. This form of criticism is more "transcendental"; cf. *ibid.*

15. For example, see 9.2.

16. The ladder metaphor comes from the early Ludwig Wittgenstein, who recognized that his own work was a kind of ladder that one had to abandon after climbing—*Tractatus Logico-Philosophicus* (London: Routledge & Kegan Paul, 1951), pp. 188-189.

17. Clive S. Lewis, *Miracles: A Preliminary Study* (London: Centenary, 1947), pp. 23-31.

18. I assume throughout this discussion that the traditional order of law-then-prophets is the correct one, and that Moses is the real author of the Pentateuch (which still allows that Moses may have used previous sources and that authorized scribal additions or clarifications might have been added here and there by those after him—e.g., Deut. 34). For discussion of this issue, see Oswald T. Allis, *The Five Books of Moses* (Philadelphia: Presbyterian and Reformed, 1969); Meredith G. Kline, *The Structure of Biblical Authority* (Grand Rapids: Eerdmans, 1972); and conservative OT introductions.

19. For an appreciation of the significance of covenant sanctions in the history of redemption, see especially Meredith G. Kline, *By Oath Consigned* (Grand Rapids: Eerdmans, 1968), pp. 39-49; Delbert R. Hillers, *Treaty-Curses and the Old Testament Prophets* (Rome: Pontifical Biblical Institute, 1964).

20. I cannot be satisfied with any hard-and-fast division between the Developmental and Culminational Preparation Periods, because the transfer of prevailing interest is a gradual one. Perhaps the most satisfactory division might be a division between Judges–II Kings on the one hand and I Chronicles–Esther on the other. I Samuel–II Kings views from an earlier, more "Kingly" viewpoint, the same events as I & II Chronicles views from a later, more Priestly viewpoint.

21. E.g., Rudolf Bultmann, *Theology of the New Testament* (New York: Charles Scribner's Sons, 1951), I, pp. 276, 279, 348, *et passim*. Doubtless there are ways of expressing these two emphases in paradoxical fashion, but they can equally be expressed nonparadoxically.

22. See John Murray, *The Imputation of Adam's Sin* (Grand Rapids: Eerdmans, 1959).

23. Note in this connection that the sabbath is used constantly as a Culminational symbol in the history of redemption.

24. Perhaps I could just as well have used the same terms, 'Prophetic,' 'Kingly,' and 'Priestly,' instead of introducing three new terms. However, I would rather reserve 'Prophetic,' 'Kingly,' and 'Priestly' for more classificatory purposes, and have a second set of terms for the purpose of "dissection," especially as this dissection applies to the history of redemption.

25. "Following the lead of the Scriptures themselves, Reformed theology has long prized the covenant as a structural concept for integrating all that God has so diversely spoken unto men of old time and in these last days"—Kline, *By Oath Consigned*, p. 13.

26. I classify God's stipulations to Adam as covenantal. Certainly the structure of covenant, if not the express word *bᵉrit*, is already present in Gen. 1: 28-30 and Gen. 2. The renewal of the Adamic mandate 1:28-30 in the time of Noah is a covenant between God on the one hand, and Noah and his descendants and the animals on the other (9:8-17). Hence presumably the original mandate also may be regarded as a covenant.

27. Note, in connection with sense (b), that the OT never uses the plural of *bᵉrit* ("covenants"). (See, however, Eph. 2:12; Rom. 9:4—but cf. variant reading.) Moreover, a later covenant is sometimes explicitly described as an establishment or fulfillment of an earlier covenant: Gen. 9:9 (cf. 6:18)); 26:3; 28:13; 35:12; Exod. 2:24; 6:4ff.; Neh. 9:32; Mal. 3:1.

28. This use of 'Axial' has nothing to do with the terms 'axiology' (4.2) and 'axiological.' The latter two terms are derived from Greek *axios* ("of like value or worth"), the former from Latin *axis* ("axle"). Metaphorically speaking, the parties are the "poles" and the Covenantal relation is the "axis" connecting the "poles."

29. Kline, *Structure*, pp. 45ff.

30. Herman Dooyeweerd and others of the "cosmonomic school" have pointed out how important "law" is to any philosophy—*A New Critique of*

Theoretical Thought (Philadelphia: Presbyterian and Reformed, 1969), I, p. 95.

31. In contrast to this, some people tend to insist on one meaning as the "right" one. For example, one receives the impression that too many writers in the cosmonomic school operate with a peculiar view of language. According to this view, a word "must" have one correct meaning, and deviations from this meaning are a sign of sin. Their polemic revolving around the words 'theology' and 'psychology' is the most obvious manifestation of this tendency. For further discussion, see 9.2.

32. Note that this is different from the cosmonomic usage of "law." So far, however, it is simply difference, not necessarily opposition.

33. The "third mode of being" language comes from H. Evan Runner, *Syllabus for Philosophy 220; the History of Ancient Philosophy* (unpublished; Grand Rapids: Calvin College, 1958–59), p. 18. See the discussion in 2.1.

34. "Although God knows whatsoever may or can come to pass upon all supposed conditions, yet hath He not decreed any thing because He foresaw it as future, or as that which would come to pass upon such conditions"—The Westminster Confession, *The Confession of Faith . . .* (Edinburgh: Free Presbyterian Church of Scotland, 1967), 3:2.

35. This is not the place to enter upon an extended discussion of the closure of the canon, which is asserted here. Arguments presented in a short compass are unlikely to be convincing to those who are not already convinced. The most important single text may well be Heb. 1:1-2, since in connection with the thrust of Hebrews as a whole it exhibits how an addition to Covenantal Words in our day would ultimately amount to a challenge to the climactic sufficiency of revelation through the Incarnate Son and through those whom he explicitly commissioned.

A related problem is that of how to identify the supposed new Covenantal Words. Throughout the OT and into the NT provision is made for the identification, collection, and preservation of additions to canon (see Kline, *Structure,* and Deut. 13; John 15:26-27). However, there is no clear indication of expectation of more Words *now,* or how to identify them. If it be said they are identified by the signs accompanying, then II Thessalonians 2:9 stands in the way. If it be said that they are identified by their conformity with Scripture's teaching, we need not attend to them anyway because Scripture already says what they say (cf. sufficiency above). The cry for more Covenantal Words does not really appreciate the difference between Accomplishment and Application of redemption (or more precisely, between the Corporate Generational Application Period and the Corporate Developmental Application Period).

36. See Appendixes 1 and 2 for some criticism of the cosmonomic view of law.

37. In other writings I have sometimes bluntly answered that the Word of God is God ("A Biblical View of Mathematics," *Foundations of Christian Scholarship: Essays in the Van Til Perspective,* ed. Gary North and Rousas Rushdoony [to appear]). I do not now say that the blunt approach is "wrong" (see to the contrary 3.32431), but I avoid it in this book because it is less precise, and open to several serious misunderstandings.

First, it is open to the misunderstanding that the Word of God cannot be other things as well. We say that Christ is God, but that does not exclude saying that Christ is man. Second, it blurs the distinction between what is personal (e.g., Scripture, God's decrees, God's faithfulness, God's righteousness, God's love), and what is a person (the Son of God). Of course, Scripture itself sometimes does the same thing (John 1:1ff.; I John 1:5; 4:8; the OT uses the "name" of God as the virtual equivalent of God).

Third, it could be understood as implying that God had to create the world and could not have decreed and declared Creation to have been otherwise than it is. But this is no more true than the claim that Christ had to become Incarnate (i.e., even apart from the fall and a free decision to save some). Fourth, it could be understood as implying that we ought to use the phrase 'word of God' in only one way. Actually, a variety of options are open, as long as we succeed in communicating the truth. We might even stop using the phrase altogether, at the cost of some circumlocution. After all, some human languages may not have a noun like 'word' but only a verb like 'speak' or 'say.' Then one would be forced to use circumlocution.

38. Or a Unit may be described as a composite whole, recognizable by observers within a system of composite wholes. The first whole has a certain amount of variation but is in contrast with other wholes. A unit has been well described when there have been specified its contrastive-identificational features, its variation, and its distribution. Cf. 3.1332, 5.21, and Kenneth Pike, "Foundations of Tagmemics—Postulates—Set I," unpublished, January, 1971, p. 9. 'Unit' is an "emic" term, that is, a term relative to a "native" observer or group of observers. A person from another culture may not recognize a Unit where the natives do.

39. Compare with this the cosmonomic language concerning things and structures "qualified by" certain modal aspects (Herman Dooyeweerd, *New Critique*, III, pp. 53ff.).

40. This corresponds vaguely to Dooyeweerd's "community." Dooyeweerd's "inter-individual and inter-communal relationships" correspond to Societal Relationships. See *New Critique*, III, pp. 177f.

41. Cf. Dooyeweerd's "institutional communities" (*New Critique*, III, pp. 187-190).

42. John Murray, *Divorce* (Philadelphia: Committee on Christian Education, Orthodox Presbyterian Church, 1953).

43. Especially in view of the artificial and sometimes damaging theories propagated under the aegis of cosmonomic philosophy. Cf. the incisive remarks by John M. Frame, *The Amsterdam Philosophy: A Preliminary Critique* (Phillipsburg, N. J.: Harmony Press, *c.* 1972), pp. 46-47).

44. Cf. Edmund P. Clowney, *The Biblical Doctrine of the Church* (unpublished; Philadelphia: Westminster Theological Seminary, *c.* 1968 [mimeographed]).

Chapter 4

AXIOLOGY

Now let us consider the third major "metaphysical" question of 1.3, namely, "Why is it there?" Like the other two questions, this one does not demand a unique answer. Like the other two questions, it may be the product of a religious malaise. One might attempt to give an obvious answer: grass is for the cattle, cattle are for milk and food, autos are for traveling, and so forth. Or one might give a simple, comprehensive answer: it pleased God to establish and work all things the way they are, for his own glory.

A more detailed answer to "Why?" could move in a number of directions. The answer could take the form of talking about what things are "good for," and in particular what human persons, deeds, and attitudes have God's approval. Under "axiology" is included not only ethics (evaluation of personal acts, motives, intentions, etc.) but evaluation of Subhuman Creatures and events involving such Creatures.

4.1 *Ethics*

Let us, however, start with human acts. Several complementary perspectives are useful in deciding what human acts have God's approval. The directives of Scripture devote some attention to (a) a more or less direct description of acts themselves, (b) motives involved, and (c) the situation in which acts or motives occur.

Hence we may speak of a "normative" or "rule" ethics in Scripture, focusing on commands: "Blessed are those whose way is blameless, who walk in the law of the Lord" (Ps. 119:1); "for neither circumcision counts for anything nor uncircumcision, but keeping the

111

commandments of God" (I Cor. 7:19). Or we may speak of a "motivational" or existential ethics in Scripture, focusing on motive: "he who loves his neighbor has fulfilled the law" (Rom. 13:8); "the fruit of the Spirit is love, joy, peace, patience, kindness, . . ." (Gal. 5:22; cf. I Cor. 13). Finally, we may speak of a "situational" ethics in Scripture, focusing on the situation and on what the results of an act will be: "so, whether you eat or drink, or whatever you do, do all to the glory of God" (I Cor. 10:31).[1]

Any one of these perspectives is, in a certain sense, sufficient to define what meets God's approval. For example, anyone who *truly* obeys the commands of God (which commands require certain inward attitudes as well as formally correct behavior) is approved by God. Likewise, anyone who truly acts from love or truly acts for the glory of God is approved by him.

Nevertheless, each perspective also presupposes the others, so that no one can operate without in effect taking all three into account. For example, Scripture commands (normative) men to love (existential) and to take the situation into account (situational). Likewise loving God (existential) involves keeping his commands (normative; John 14:15; I John 5:3), and desiring his glory (situational). Similarly, the glory of God is always served by keeping his commands and loving him (John 15:8). In sum, there is a proper, biblical deontological (normative) or rule ethics, an existential or personal or motivational ethics, and a teleological or situational ethics. But these three perspectives are not in tension or in competition as they are in non-Christian versions of ethics. The "situation" cannot be played off against the rules as Joseph Fletcher does;[2] on the contrary, the rules are *part* of the situation established by God.

> Description. The *Normative, Existential,* and *Situational Perspectives* on Ethics are ways of looking at ethics which focus respectively, on (a) commands, (b) persons and their motives, (c) the situation.

4.2 *Axiology in general*

The above threefold division can also be generalized to axiology as a whole.

Description. *Axiology* is the study of value.

Description. The *Normative, Existential,* and *Situational Perspectives* on Axiology are ways of looking at the value of Items which focus, respectively, on (a) rules concerning value (whether God's or Men's), (b) the Items themselves in their dynamic development, (c) the situation in which the Items occur.

Ethics as a subdivision of Axiology could be described as follows.

Description. *Anthropological* Axiology is the Axiology concerning Men's persons, deeds, intentions, and dispositions. (This is in contrast to Proper-Theological, Angelic, or Subhuman Axiology.)

Description. *Ethics* is Sanctional Anthropological Axiology, that is, the study of what human persons, deeds, intentions, and dispositions warrant approval.

Note: this means that Ethics is in contrast to study of what *interests* human persons, deeds, etc., may serve even though they may *not* warrant approval. For example, using the First Polar View, Locutionarily the Creation remains "good" even after the fall (I Tim. 4:3). The bodies of unbelievers are not in themselves unpure. Likewise, Administratively unbelievers serve God's purposes willy-nilly, whether they wish to or not (Prov. 16:4; Ps. 76:10). But Sanctionally, they are under his curse (Gal. 3:10; Ps. 5:5). Ethics is primarily concerned with this third element, namely, God's Sanction on human persons, deeds, etc.

Ethics can be subdivided with reference to the Dominical, Covenantal, and Servient Bond.

Description. Dominical, Covenantal, and Servient Ethics are, respectively, the Ethics of the Dominical, Covenantal, and Servient Locutions; that is, the Ethics that answers the questions what does God approve, what does he say in the Covenant that he approves, and what do men approve.

Dominical Ethics and Covenantal Ethics almost coincide, since the Bible is sufficient to equip the man of God for every good work (II Tim. 3:17). The man who keeps Covenantal law is blameless (Ps. 119:1). In other words, Scripture gives us all that we need to know about what human persons, deeds, etc., God approves. However, Dominical Ethics can still include many unanswerable questions about (say) whether God approves such-and-such past action. Frequently our knowledge of people's motives and circumstances (as well as our lack of wisdom) does not permit a definitive evaluation. Ethics can be subdivided in terms of Normative, Existential, and Situational Perspectives, much as we did above. The same subdivision can be described in terms of the Prophetic, Kingly, and Priestly Functions. Approved persons, deeds, and intentions can be described as either (a) those in accordance with God's commands (Prophetic), (b) those that are the working of God's redemption (Kingly; cf. Phil. 2:13; Ps. 119:68; James 1:17), or (c) those that receive God's blessing (Priestly). These three are the Normative, Existential, and Situational Perspectives, described this time from a First Polar rather than a Second Polar View. (In the Second Polar View we describe a *man's* obeying commands, working with proper motives, and working for God's glory.)

The correlation between Normative/Existential/Situational and Prophetic/Kingly/Priestly confirms the observations in 3.35 about the relation of both these sets of categories to Particle /Wave/Field.

4.3 Axiology in relation to ontology and methodology

Section 3.34 on interlocking of categories can now be expanded to include a discussion of the interlocking between axiology and other categories. Discussion will for the most part be confined to a short sketch.

4.31 Axiology applied to ontology

Any of the four Kingdoms can be subdivided in terms of its usefulness ("value") to God, to man, and to various Subhuman Creatures.

4.32 Axiology applied to methodology

We could refine the discussions of modality, temporality, or structurality by taking value into account. To some extent this has already been done in our previous discussion (see, for example, the discussion of reductionism in 3.133, or the fall under 3.2).

4.33 *Ontology applied to axiology*

We have already begun this application with the distinction of Anthropological Axiology from (say) Subhuman Axiology.

4.34 *Methodology applied to axiology*

The Normative/Existential/Situational distinction is an application of the Prophetic/Kingly/Priestly distinction to Axiology. The Dominical/Covenantal/Servient distinction from structurality has also been used. Temporal categories could be applied by discussing the change and development of values in various Periods.

Chapter 6 will bring these Ethical considerations to bear on science.

NOTES TO CHAPTER 4

1. For my insight into these three types of ethics—and for my terminology as well—I am indebted to classroom lectures by John M. Frame, Westminster Theological Seminary, Spring, 1973. Mr. Frame in turn has built on Cornelius Van Til, *Christian Theistic Ethics,* In Defense of Biblical Christianity (Philadelphia: den Dulk Christian Foundation, 1971), III.

2. Joseph Fletcher, *Situation Ethics; the New Morality* (Philadelphia: Westminster, 1966). Fletcher says, for example, "It [situation ethics] goes part of the way with Scriptural law by accepting revelation as the source of the norm while rejecting all "revealed" norms or laws but the one command—to love God in the neighbor" (p. 26).

Chapter 5

EPISTEMOLOGY

How do we come to know what we know? What does it mean that a person knows something? What is knowledge? What is knowledge knowledge *of?* Because of the intimate involvement of science with knowledge, such questions cannot be neglected by philosophy of science. By asking them we can come a step closer to dealing with modern science from a biblical point of view.

The above questions, like the questions of 1.3, are questions with some mystery in them. Just what is it that the inquirer wishes to know? If the inquirer asks, "How do you come to know that the treasure is buried there?" there is no mystery. One answers, "Because I buried it there myself," or "Because I found the map." If, on the other hand, he asks, "How do you come to know in *general?*" it is not obvious what answer would satisfy him. The trouble is that there are many different types of knowledge, and many different ways of coming to know things that you know. One knows about mathematics because he studied it in school, one knows about dogs because he has raised some, one knows about the accident because he was an eyewitness, one knows that inflation is coming because the government is increasing the money supply.

If, however, we assume that the question is posed because of a religious malaise as to how we can know anything, a biblical answer can be given in terms of man's relation to God. A man comes to know something as God reveals it to him. A man knows about something when he grasps some of the truth about it, the truth ordained by God. Knowledge includes and involves knowledge of God and Creation (ontology), knowledge of how to do things and how things

116

function (methodology), and knowledge of the truth about things (axiology).

Now let us look more closely at knowledge in relation to (a) God, (b) what is known (the "subject-matter"), and (c) human knowers.

5.1 Knowledge in relation to God

I will discuss knowledge in terms of what is known (ontologically: 5.11), how it is known (methodologically: 5.12), and the value of knowledge: 5:13).

5.11 Ontologically

God knows all things (I John 3:20).[1] This includes a complete knowledge of himself (Matt. 11:26-27; I Cor. 2:10), and of the past, present, and future of the Creation. Some have denied that God knows the future in exhaustive detail before it happens. (Usually they are under the illusion that this will open the door to human autonomy.) But in view of Lamentations 3:37; Ephesians 1:11; and Hebrews 1:3, this involves a denial that God knows what he himself will do. Hence the denial amounts not really to an attack on God's sovereignty so much as on God's decision-making ability—which result is very far from what the deniers usually desire. Moreover, the whole scheme sets a limit to God's understanding, contrary to Isaiah 40:28 and Psalm 147:5.

God's knowledge can be further described in terms of Prophetic, Kingly, and Priestly Functions. God knows the truth about everything because (a) he has ordained this truth (Prophetic), (b) he *works* all things according to the counsel of his will (Kingly), and (c) he is himself the standard in terms of which truth and falsehood are evaluated (Priestly).

Second, Created persons, both angels and men, know things. Since our knowledge of angels is slim, I will confine my discussion to men. Men know God (Rom. 1:21), they know something of the law of God (Rom. 2:14-15), they know about affairs of everyday life (II Sam. 11:16; 14:1, etc.). Hence they know *some* of what God knows. But let us pass on to consider how persons come to know what they know.

5.12 *Methodologically*

God is the origin of knowledge. There is no time when he did not know everything. Hence it would be misleading to speak of God "coming to know" something. Nevertheless, the Bible does tell us something of how God knows what he knows. He knows because he created what he knows (Active; Ps. 33:15), because he is everywhere present and in all things (Middle; Jer. 23:24), because he observes all things (Passive; Ps. 33:13-15).

Now let us consider Men. Knowledge is one of the gifts that God has given and continues to give to men. He teaches men knowledge (Ps. 94:10) and makes them understand (Job 32:8). The Bible speaks especially of the fact that all men know God, as a result of God's showing what he is like to them (Rom. 1:19-21). Through the Holy Spirit believers are taught to know God and his will in a saving way (I Cor. 2:10ff.; I John 2:20-21, 27). Paul also mentions a special gift of knowledge (I Cor. 12:8). This might mean a gift of being able to know things by Covenantal Locution[2] that are not ordinarily knowable—as when Agabus predicts a famine (Acts 11:28). More likely, it has to do with the intensification and deepening of the kind of knowledge that every believer has. When the other lists of gifts and offices are compared with I Corinthians 12:8-10, the gifts of wisdom and knowledge are seen to be closely related to the prophets and teachers of I Corinthians 12:29, to the teaching offices in Ephesians 4:11, and to the prophecy, teaching, and exhortation of Romans 12:6-8. Thus it is clear that the knowledge of Christ and his redemption is a gift of God.

Next, what about knowledge less intimately related to salvation? What about knowing that a book is on the table? The focus of Scripture is on the knowledge of Christ. Anyone who does not know him is a fool (Ps. 14:1; Rom. 1:21-22; Eph. 4:17-19). But Scripture doubtless implies that even the knowledge that a book is on the table is a gift of God and has been shown to us by God. God gives us our food and clothing (Luke 12:22ff.; Acts 14:17). Can it be denied, then, that he gives knowledge in general? It would appear not, especially in view of the general statements in James 1:17

and Romans 11:36. Furthermore, there seems to be no adequate reason for restricting the scope of Job 32:8-9 and Psalm 94:10.

God's revelation to man can be further analyzed in terms of Prophetic, Kingly, and Priestly Functions. God tells to men, or causes to be told to man (cf. Ps. 19:1ff.), what he knows (Prophetic); God *empowers* man to know (Kingly); and God *blesses* man with the gift of knowledge (Priestly). Or, again, we can look at God's revelalation to man from Generational, Developmental, and Culminational Views. God puts man in a situation to learn and know what he knows (Culminational); God causes man to know (Developmental); and God enables him to use and apply what he knows (Generational).

5.13 *Axiologically*

Under the heading of axiology come topics like the validity of claims to knowledge (normative), the responsibilities involved in having knowledge (existential), and the value of such knowledge as a person possesses (situational?).

First, a word about God's own knowledge. As we have seen, God claims to know everything, and more specifically he claims to know what he has Covenantally revealed. In various ways God confirms his claims and "makes them good": Isaiah 41:26ff.; Ezra 1:1ff.; Acts 13:33, etc. However, even when the evidence seems to go against his claims, we are bound to believe them, because he is the Lord (cf. Rom. 4:18-21; Ps. 77).

God's unfathomable knowledge also carries with it a "responsibility" on his part to use his knowledge rightly. Yet since God is infinitely wise, and is himself the standard for responsibility, the one to *whom* men are responsible, it is impossible that he should himself fail to be "responsible." Finally, we may say that there is great value in the knowledge that God has. He finds satisfaction in it, as he finds satisfaction in all his works. This applies especially, of course, to the Father's knowledge of the Son and the Son's of the Father, since it is a knowledge of love.

Second, consider the knowledge that men have. Here let us distinguish Normative, Existential, and Situational Perspectives.

5.131 *The Normative Perspective on men's knowledge*

The standard for men's knowledge is God's knowledge, inasmuch as their knowledge is derivative from his. Not only does he give men whatever knowledge they know, but whatever they know agrees with what he knows.

As a result, claims to knowledge by men cannot be accepted in the same way as claims by God. Claims may be compared with what God says in the Bible. No claim in competition with or against what God says *ought* to be accepted.

5.132 *The Existential Perspective on men's knowledge*

All men's knowledge involves a personal relationship to God. This is so, first of all because men are responsible to God to thank him for giving them the equipment making it possible to know. They are responsible also to thank him for giving them knowledge, and for empowering them to retain the knowledge that they have. They are further responsible to him for properly using what they know.

But that is not all. In knowing, for example, that a book is on the table a man at the same time knows something about God. Romans 1:19-21 expresses this in general by saying that God shows himself (*ephanerōsen*) in all Creatures (*tois poiēmasin*), and this certainly includes his showing himself in books and tables. "In him we live and move and have our being" (Acts 17:28). In particular, a man knows something about the fact that God has ordained that the book be on the table, that God causes the book to be on the table, that he is pleased to have the book on the table.

A man may, of course, evade his responsibilities to God in several interrelated ways. Romans 1:21ff. anticipates and describes this. He may make for himself an idol to which he diverts his responsibility, or he may deny that there is any God. Either one of these moves involves a certain attempt to enthrone himself as God—to claim human autonomy. This is the "ontological" side of idolatry. Second, a man may *methodologically* try to eliminate reference to God from all his knowing, to deny that knowledge that a book is on the table involves knowledge of God, and to claim that there is no need for God in epistemology. Third, a man may *axiologically* refuse obedience and submission to God, even though he knows that God shows himself in all knowledge of the truth. The devils and Satanists are perhaps most representative of this side of rebellion.

However, these three rebellious moves *are* interrelated, so that none takes place without some degree and form of the others. For example, the refusal of obedience and submission to God (axiological) most often takes the form of denial of the personal character of God (ontological). If the "being" who gives men knowledge and causes the Creation to be what it is is not personal, it leaves a man free from personal responsibility for obedience and thankfulness to this being. Hence the attractiveness of the images like birds, beasts, and creeping things (Rom. 1:23). Naming God "the Absolute" or "the First Cause" or "Nature" or "Chance" or "Natural Law" or "Being-in-Itself" has similar effect. Similarly, if methodologically one claims that God is not really all *that* involved in the world, one's responsibility to God becomes remote and irrelevant (axiology).

It is ironic that each of these rebellious "moves" exploits powers of man and facts about this Creation that God has ordained. It is another case of Van Til's dictum that the father's small child must sit on his lap in order to slap him.[3] Rebellious man must use God's gifts in order to insult God.

For example, in ontological rebellion a man ascribes to Creatures or to the "Absolute," etc., characteristics that God has. Thereby he admits that he needs God, and yet he will not accept God with all his characteristics. Moreover, in ascribing such characteristics to a *Creature* he exploits the fact that Creatures themselves, as God's handiwork, display God's power and deity. Idolatry is parasitic on so-called "general revelation."[4]

Second, methodological rebellion depends upon man's (created) ability to focus on one or more Items, leaving others in the background (cf. 3.133). The fact that we can talk about knowledge and what we know without specifically mentioning God gives plausibility to the claim that God can be *eliminated* as a factor in knowing that a book is on the table. But not talking about God is something like not talking about the air around us. If we do not need to mention it, that is only because it is *always* there.

Thirdly, axiological rebellion depends upon the fact that a man does not have responsibility toward the Subhuman Kingdom or toward ideas ("the Absolute") in the way that he does toward persons. Moreover, a man's responsibility to other human persons is limited. Hence if God is treated in a way analogous to how man has been taught (by God!) to treat Subhuman Creatures or Men, obedience can be largely evaded.

5.133 *The Situational Perspective on men's knowledge*

A man ought to use his knowledge for the service of God's kingdom. To "really" know, he must know what his knowledge is good for. For example, suppose a man has been trained by rote to count from 1 to 10, and to write addition problems correctly in the form "$2+5 = 7$," "$3+1 = 4$," etc. Suppose now that we show him four apples, and ask how many there are, or ask how many there would be if we added two more—and suppose that he fails completely to fathom our meaning. Would we say that he "knows" that $4+2 = 6$? Has he not rather learned how to play a meaningless game in which certain composite symbols are acceptable ('$2+5 = 7$,' '$3+1 = 4$') and others are not ('$2+1 = 4$')? He may not have the foggiest idea what 'two' and 'four' mean in English. Thus knowing that $4+2 = 6$ involves knowing more than just how to mouth words; it involves an ability to do certain things in practical cases (as with the apples).

Thus in a certain sense we may say that an unbeliever does not "know" anything the way a believer does—because the unbeliever does not know what it is good for. He does not have the ability to use anything for God's kingdom and service.

Yes, an unbeliever can know that a book is on the table. He picks the book up, rather than trying to make it go down. He does not expect the table to be penetrable or the book to explode, showing that he knows something about tables and books. So in the "short run," as it were, he can manage pretty well, feeding himself, protecting himself from danger, and talking coherently about the book and the table.[5] But in the "long run," he fails miserably, because nothing he does is done in true service to God. So, viewed from a larger perspective, he does *not* know what it means that the book is on the table. He does not use this truth in the way that it is asking to be used. This is one reason why the Bible speaks again and again of the ignorance of unbelievers (e.g., Eph. 4:17-19; Gal. 4:8).

5.2 Knowledge in relation to subject-matter

Knowledge is both knowledge about things, knowledge of how to do things, and knowledge that certain things are true. For instance, we say, "I know the way to the city," "I know my dog," "I know about dogs" (knowing about), "I know how to repair an automobile" (knowing how), "I know that he will arrive tonight" (knowing that). These three ways of talking about knowing are, of course, not rigidly separated from one another. Knowing *that* he will arrive tonight involves knowing something *about* him (e.g., that he is a person or at least an animal of the male sex, not a plant or Inorganic Creature or a quality) and knowing, to a certain extent, *how* to deal with the matter of his arriving.

5.21 Knowing about

Let us first focus on "knowing about." Knowing about a Unit involves an ability and appreciation concerning its Contrast, Variation, and Distribution.

> Description. The *Contrast* of an Item involves those features that identify it and contrast it with other Items. "Items which are independently, consistently different are in contrast."[6]
>
> Description. The *Variation* of an Item is the range of difference through which it may vary while still remaining recognizably the "same." (Obviously this depends on what kind of "sameness" we may be interested in.)
>
> Description. The *Distribution* of an Item is comprised by the neighborhoods in which it may occur. This may be further analyzed into Distribution in class (Particle), in sequence or location (Wave), and in system (Field).[7]

The Contrast/Variation/Distribution distinction is intimately related to the Particle/Wave/Field distinction, as this latter distinction applies to Units. For example, knowing one's brother involves appreciating his Contrast with other family members, with other persons, and with Subhuman Creatures. He is, say 6'0'', has wavy red hair, is interested in chemistry, etc. It also involves being able to identify him through all the changes that he undergoes. Has he just had a haircut, or taken off his glasses, or eaten a hamburger? These are Variations in one's brother.

Moreover, one should appreciate how the brother is Distributed in class: the class of family members, of males, or persons. He is Distributed in sequence and location: born at a certain time, destined (if the Lord tarries) to die, living in a certain city (not on the moon or at the bottom of the sea). He is Distributed in the system of persons as that class can be broken up in terms of occupation, age, sex, hobbies, education, height, etc.[8]

Some of the anti-Christian philosophical problems of epistemology can be viewed (at least in part) as attempts to deal with the question of knowledge in terms of one or at most two of the CVD triple above. For example, the idea that knowledge is only of universals (and hence that we cannot know our brothers, but only brother-ness or wavy-haired-ness) becomes plausible by focusing only on Contrast. The Contrastive features of Units (having wavy hair, a certain height, etc.) are somewhat like so-called "universals."

On the other hand, the idea that knowledge is only of particulars (atomism or nominalism) arises from focusing on Variation. One's brother *is* different from day to day, and we notice differences. The mistake is then to say that we know brother-instances but *not* our brothers. We have already seen in chapter 1 that there are Creatures; hence it is unbiblical to imagine that one eliminates Creatures by observing or focusing on Creature-instances (by focusing on the Variation of Creatures).

Moreover, the early Wittgenstein[9] may be taken as an example of an almost exclusive emphasis on Distribution. The world consists of facts (not things). However, an element of nominalism remains in his starting with atomic facts.

5.22 *Knowing how*

About "knowing how" one could say much the same as we said about "knowing about." Contrast, Variation, and Distribution are naturally involved in both. However, it is also interesting to bring into play more directly some of the material developed in chapter 3. "Knowing how" involves components from a Particle, Wave, and Field View. Knowing how to repair an automobile, for example, involves having an idea of how to distinguish what is wrong, and what is the difference between proper and improper functioning (hence modality) of the parts. It involves also being able to plan, execute, and complete a temporal program of repair (temporality). Finally, it involves coordinating the parts with one another and with the performance desires of the owner (structurality).

5.23 Knowing that

One can also analyze the facts *that* one knows in terms of Contrast, Variation, and Distribution. For example, one knows what $2+2 = 4$ means in terms of (a) its Contrast with other knowledge, that $2+3 = 5$, $2+4 = 6$, etc., and falsehoods $2+2 = 5$, $2+2 = 6$, etc. (b) One appreciates the Variation among situations where $2+2 = 4$ applies: four apples, four houses, four people, etc. (c) One recognizes $2+2 = 4$ as Distributed in class, sequence, and system. "$2+2 = 4$" is in a class of numerical equational truths, it occurs in sequence in certain linguistic and nonlinguistic contexts having to do with numbers of things, and it occurs in a whole system of numerical calculation with whole numbers, fractions, and decimals, with paper, slide rule, or adding machine, and so on.

5.3 Knowledge in relation to men

We will discuss knowledge successively with ontological, methodological, and axiological emphasis.

5.31 Ontologically

Men know things, both individually and in groups. To some degree they share knowledge with one another, both because of similar experience, similar opportunities, and similar communication. We know some things in part because they have been handed on from previous generations, and we in turn are engaged in teaching our children and grandchildren. It might even be said (though perhaps this stretches vocabulary a little) that a group may know how to do things that no one individual in the group knows how to do.

The church too has a tradition from which the individual believer ought not to separate himself. To use the biblical metaphor, he as a member of the body does not by himself make up an entire body. He needs the strength, the blessings, the insights that other members can give. In particular, since we are talking about knowledge, he needs to respect the gifts of wisdom, of knowledge, and of teaching that other members have (I Cor. 12:8, 28). Since the Holy Spirit did not begin to give gifts to the church just yesterday, this involves also listening to the great teachers of the church from the past. Of course, this does not mean slavish acceptance (except of the apostles' and inspired prophets' writings). It means receiving believers' teachings as far as they faithfully expound and interpret the Bible.

5.32 Methodologically

Knowing belongs to the Cognitional Function. But, as usual, because of the interlocking of Functions, we cannot go very far without realizing that all the Functions are involved in Cosmic Men's knowing. I have already, in 5.31, pointed out some ways in which the Social Function is involved. Behavioral, Biotic, and Physical Modes are involved in quite basic ways, since a man, in order to know about the Cosmos, must retain a Physical connection with it (which he loses at death), and he must be alive (Biotic) and be able to perceive things (Behavioral).

Next, look at knowing in terms of the Prophetic, Kingly, and Priestly Functions of man. The Cognitional Function itself is an aspect of the Prophetic Function. And it cannot be denied that knowledge is intimately connected with language. Not only do men communicate knowledge to one another (5.31), but even in individual reflection we frequently use verbal or other symbols without utterance out loud. God has given men the gift of language as part of the equipment for knowing about God's Cosmos.

Moreover, especially in the "knowing how" side of knowledge, man's knowledge involves a certain power to influence the Cosmos in certain ways—thus exercising a Kingly Function. Finally, a man must be able to discern purpose (grass for cattle, cows for milk, etc.) and to use what he knows with purpose. He is active in a Priestly way in the appreciation that knowing involves.

Because of the varying interests of men, knowledge can be specialized in various ways. Under sections 2 and 3.1 we have already seen how that we can talk about study focused on a specialized subject-matter. On the other hand, knowledge can also have a focus of being adapted for communication to men. Or knowledge can focus on how what one knows relates to God.

> Description. *Sensitive* knowledge is knowledge specialized with respect to subject-matter. *Refined* knowledge is knowledge adapted for communication to men. *Sapiential* knowledge is knowledge involving spelling out some of the relations of what one knows to God.[10]

For example, suppose that we come across a nest of ants. Acquiring Sensitive knowledge of the ants means exploring how the ants have built this nest, how they function in it, how they are reproducing, etc.—without necessarily worrying about whether we can communicate this knowledge to another person, or convince him of the same truths. Sapiential knowledge would reflect on God's power and purposes with ants and with this particular nest. We might be led to praise God for the marvel of his wisdom displayed in the nest.

Finally, acquiring Refined knowledge would involve exercising care that we seek the type of knowledge interesting or useful to others also, and acquiring it in such a way that people (oneself included) are really convinced. Refined knowledge typically involves an effort at (a) generality of scope (ontology), (b) a consistency, repeatability, and appropriateness of method in exploring the subject-matter (methodology), and (c) attention to grounds and justification of results (axiology).

Thus, for example, Refined knowledge of the ant nest might involve exploring ant nests *in general,* or what one can say about *all* ants in the nest. The public at large, after all, is not interested in *this* nest so much as in nests like it, or ants like it, that might appear at other times and places. Or, methodologically, Refined knowledge might involve developing methods for observing nests with a minimum and predictable disturbance of the inhabitants, and of taking statistical samples of ants or ant nests. This would again be with the effort to avoid radically changing the situation when *another* person disturbs the nest, or makes some numerical estimates. Axiologically, obtaining Refined knowledge might involve the introduction of special hypotheses and their testing, the recording of stages in the results, etc., so that another person can see the train of analysis that led to the knowledge in question.

5.33 *Axiologically*

Now let us deal more specifically with the effects of the fall of men on knowledge. Men are either covenant-keepers or covenant-breakers. That is, they have a fundamental (Sapiential) orientation either toward serving God, or toward rebellion against him (I John 2:29–3:20; Rom. 8:5ff.). They are members of God's kingdom or they are not (Col. 1:13; Gal. 1:4).[11] However, no covenant-keepers in the Cosmos, except Jesus Christ in his Cosmic life, are free from sin (I John 1:8ff.). Hence an analysis of the way that sin affects men's knowledge will be relevant to both Cosmic covenant-keepers and covenant-breakers, though it will be characteristic of covenant-breakers.

I propose now to go through each of the headings in 5.11–5.32, specifically focusing on what light these sections can shed on the effects of sin on knowledge.

5.3311 *Some implications of 5.11 for sin*

God knows everything. When a man knows something, he knows something that God knows. If he has constructed for himself an idol, he will tend to allow and admit only such knowledge as his idol could "know." For example, if his idol is "modern science," then the only knowledge is "knowledge" as "modern science" has. (In particular, he cannot know his wife.) If his idol is humanism, he will accept the "knowledge" about which there is a consensus in Western humanism.

5.3312 *Implications of 5.12 for sin*

God gives men all the knowledge that they have. But if God is angry with a man, he may withhold knowledge (Rom. 1:21-24, 28). Moreover, if a man is angry with God or does not trust God, he will be inclined to disbelieve what God teaches him, and hence his knowledge will be impoverished. This is true of all knowledge, but it is more obviously true of the truth in the Bible. If a man does not believe the Bible, he knows less.

Second, if a man has mistaken notions about God, if he has made idols, then he will be less able to identify true teaching, that is, teaching from God, and will be apt to receive teachings of demons (I Tim. 4:1). Third, since all revelation from God has a Personal Prophetic/ Kingly/Priestly structure, the attempt to deny God can be carried forth in a thorough-going way only by a denial of the meaning of knowledge and persons.

5.33131 Implications of 5.131 for sin

If a man does not properly distinguish between the claims of God and the claims of men (e.g., by introducing extrabiblical "revelations" like the Book of Mormon or by denying the divine authority of the Bible), then he believes some things on insufficient evidence and rejects others even though the evidence is sufficient.

5.33132 Implications of 5.132 for sin

Many of the implications have already been spelled out in 5.132 itself. The fact is that men may cease to seek knowledge or seek only certain kinds of knowledge because knowledge involves personal relationship to God, and they want to avoid such relationship to God.

5.33133 Implications of 5.133 for sin

If a man loses the perspective of what it means to serve God's kingdom, to that degree his knowledge is distorted. He no longer knows how to use what he knows. Even here, distortion cannot be complete, because covenant-breakers live in God's Cosmos. They cannot completely escape their conscience and a knowledge of God's demands (Rom. 2:14f.).

5.3321 Implications of 5.21 for sin

Knowledge of things involves acquaintance with their Contrast,

Variation, and Distribution. Impoverishment obviously occurs if, as in realism, nominalism or the early Wittgenstein, one or more of these three is neglected. However, sin also affects how a man is acquainted with the Contrast, Variation, and Distribution of a Unit. For example, let us take a man's knowledge of Newton's laws of motion.[12]

Contrast. If a man sinfully holds to a reductionism (3.133), then he will be disposed to ignore or misdescribe contrastive-identificational features Weighted in some Functions. Newton's laws are likely to be thought of as *only* Energetic, or only Mathematical, or as unrelated to the Economic, Lingual, etc. In fact, the laws have a practical purpose of enabling man to make certain calculations (Lingual) and predictions (Prophetic?), which in turn will be for his Economic benefit.

Of course, in practice few would deny that Newton's laws can be so used. But the person who construes Newton's laws as a kind of metaphysical ultimate explanation, showing that the world is "only" a deterministic clockwork of material atoms, is naturally disposed to de-emphasize such interconnections. For him what Newton's laws "really" mean is materialistic. Exclusive Reductionism is visible here.

Moreover, Newton's laws of motion *contrast* with other possible patterns of order or disorder that one might have met in the motion of Inorganic Creatures. On the basis of a few observations about motions, various kinds of possible "simple" laws of motion might be suggested. Ptolemy's, Aristotle's, and Newton's ideas of what was the "simplest" law were different. These postulated "simplest" laws are basically guesses about what God has ordained. And the type of guess that one makes will be influenced by his knowledge of God.

Nor does the problem of making "guesses" disappear after Newton. Albert Einstein, for example, objected to quantum mechanics as an ultimate explanation, because it would involve believing in "God playing dice."[13]

Variation. Estimates of how far one's knowledge extends and how precise one's knowledge is are interlocked with one's estimates of how far the plans of God are predictable. The believer, for example, supposes that Newton's laws apply (though not with infinite exactness) to ordinary motions from the Adamic Period to the second coming of Christ—with possible obvious exceptions of Noah's flood, Joshua's long day, and some other miracles.[14] Most unbelievers, by contrast, would generally claim that "physical laws" hold at all times and places. Hence a difference in estimate of the Variation involved in Newton's laws is a straightforward result of attitude toward God.

Distribution. First, in terms of Distribution in class, the possible options for laws of motion may be influenced by whether one believes that "anything is possible" (it is a chance world), or that the human mind must always be competent to discern the laws, or that laws are laws ordained by God. Since Newton's laws are Distributed in the class of possible laws, this influences one's view of Newton's laws.

Second, in terms of Distribution in sequence. The environment in which one meets Newton's laws includes rules of thumb for applying the laws to various Physically Weighted situations, rules for interpreting the meaning of the symbols in the equations, and rules for simplifying situations in order to be able to apply the laws to them (e.g., represent bodies by point masses). The degree to which such environment is even noticed will partly depend upon whether one wants to press for a Physical Reductionism. Moreover, one aspect of Distribution in sequence is the tracing of the epistemological origins and applications of the laws. A balanced view of origins will acknowledge that the discovery of such "laws" is due to God, and their application must be in conformity with the will of God.

This leads us to Distribution in system. Newton's laws are Distributed in the system of modern Physics, and Physics in the system of sciences. Again, Reductionism will affect one's account of the system of sciences. More important, physics is Distributed in the system composed of the community of physicists, and physicists will be improperly understood apart from the Bible's teaching on the nature of man, and of the purposes for studying Physics.

5.3322 Implications of 5.22 for sin

Particle, Wave, and Field Views are involved in knowledge. Newton's laws of motion approach motion primarily from a Particle View.

A denial of the reality of the other two views is involved when Newton's laws are presumed to be a complete account, as in mechanism. Mechanistic reduction is related to a lack of appreciation for the richness of God's wisdom.

5.3323 *Implications of 5.23 for sin*

We have already discussed the implications of Contrast, Variation, and Distribution in 5.3321.

5.3331 *Implications of 5.31 for sin*

A person's appreciation for Newton's laws is highly influenced by the community of scientists with which he is in communication. For example, contemporary students of physics have a very different attitude toward Newton's laws than did students of the nineteenth century, even though the students themselves may never have performed experiments revealing relativity or quantum effects. They take the word of their professors. In such ways sinful or righteous thinking, either one, can be passed on.

Some people, of course, begin to realize the degree to which their knowledge and supposed knowledge depend on taking the word of their predecessors. Frequently, they are tempted to adopt Descartes's method of doubt in order to avoid the problem. But (a) Descartes's method *itself* is part of the heritage that they have received from the past, and they are not willing to doubt that part of their heritage. (b) Methodological doubt of the claims of Scripture, that is, of the claims of God, is sinful. (c) Even methodological doubt of human authorities can be of the sort that is ungrateful and disrespectful to God for the gifts that he has given men in the past. A desire to establish knowledge from the ground up, all by oneself, is essentially a conceited desire.

5.3332 *Implications of 5.32 for sin*

All coming to know on man's part involves some exercise of his Prophetic, Kingly, and Priestly abilities received from God. He is able to know about the Creation because Creation itself is the work of God as Prophet, King, and Priest. In exploring Creation man is, as it were, following in God's footsteps. Hence when he denies God he can no longer account for why he should be able to know anything using these "abilities" (or are they disabilities?—how can he tell?).

5.3333 Conclusions

Though in 5.33 I have spoken in terms of "knowledge," I could equally well have spoken about language, or about thought, or even about words. Even our knowledge of how to use words involves appreciation for their Contrast (with other words—in meaning, form, and structure), Variation (a given word can be used in a variety of ways, with a certain range of meaning), and Distribution (a word is a verb, noun, modifier, etc.).[15]

It should be clear that sin affects all our knowing—sometimes subtly, sometimes boldly, sometimes in "small" ways, sometimes in earth-shaking ways. This is true of both Sensitive, Refined, and Sapiential knowledge. Thus what we are doing here is, if you like, a critique of thought, not merely a critique of "theoretical thought."[16]

NOTES TO CHAPTER 5

1. Cf. Acts 15:17-18. The preferred variant for Acts 15:18 is *gnōsta ap' aiōnos*, in which case *gnōsta* would refer, at least primarily, to things prophesied in the OT. The implied subject of 'know' would then be men (rather than God), and *ap' aiōnos* would mean "from of old" rather than "from all eternity." Hence this text ought not to be pressed into service as an immediate demonstration of God's omniscience. It is, however, one of many texts that illustrate God's ability to prophesy the future (cf. Isa. 41:22-23; 44:6-8). And if God can infallibly prophesy men's actions in one case, why should we deny that he knows the future in all cases?
2. The ordinary name for this in systematic theology is "special revelation."
3. Cornelius Van Til, *Jerusalem and Athens*, ed. E. R. Geehan (Philadelphia: Presbyterian and Reformed, 1971), p. 403.
4. "General revelation" is the usual systematic-theological term for what I would call teaching from Dominical non-Covenantal acts of God in the Cosmos.
5. Even these abilities are gifts from God, not common to the whole human race. Babies, imbeciles, and the mentally deranged are exceptions. The abilities are good—as can be inferred from their relation to the original constitution of man before the fall. Yet the fact that a man has received such abilities from God does not weigh in favor of the man's approval in God's sight, any more than the lack of abilities weighs in his disfavor. Cf. Luke 12:47f.
6. Kenneth L. Pike, "Foundations of Tagmemics—Postulates—Set I," (unpublished; January 8, 1971 [mimeographed]), p. 9.

7. *Ibid.*, p. 5. For greater clarity readers should consult this paper, which also attempts descripitions -of words like 'independently,' 'consistently,' 'class,' and 'system,' all of which are used in a somewhat special sense. 'Contrast,' 'variation,' and 'distribution' are the present-day names for what Pike earlier called the feature mode, the manifestation mode, and the distribution mode— Kenneth L. Pike, *Language in Relation to a Unified Theory of the Structure of Human Behavior*, 2nd rev. ed. (The Hague-Paris: Mouton, 1967), pp. 84-93.

8. For further examples, see Pike, *Language*, especially chapt. 3.

9. Ludwig Wittgenstein, *Tractatus Logico-Philosophicus* (London: Routledge & Kegan Paul, 1961). One may pose the question whether the later Wittgenstein, despite all differences, does not equally have a one-sidedly Distributionist method.

10. Sensitive knowledge, especially sensitive knowledge with a focus on individual Creatures, is somewhat like cosmonomism's knowledge belonging to "naïve experience." Refined knowledge is somewhat like cosmonomism's "theoretical thought." Sapiential knowledge does not correspond *well* to anything in cosmonomism, but a cosmonomic "knowledge belonging to the heart" would perhaps be the closest analogue. However, these comparisons should be taken as a suggestion of some vague relationship, rather than as an exact correspondence. My terms are intended not to be precise, but to be useful in marking out overlapping, interlocking areas of knowledge. This book, for example, is first of all Sapiential, but also has a considerable degree of Refinedness.

11. Much of the NT language quoted here applies most specifically and Pointedly to the Application Period (the resurrection of Christ, for example, being presupposed in Rom. 8:9; Gal. 1:4, etc.). Yet a preliminary, anticipatory form of the distinction occurs in the difference between covenant-keepers and covenant-breakers in the OT. For the sake of brevity, I will not here discuss as a separate question the knowledge of men in the Preparation and Accomplishment Periods.

12. Of course, Newton's "laws" are not Laws in the sense of 3.3241. Rather, they are a human approximate description and "guess" at what God's Word is for motion during the time from creation to consummation. I use the term 'laws' here in conformity with popular usage.

13. Quoted from a letter (7 November 1944) to Max Born, in Max Born, "Einstein's Statistical Theories," *Albert Einstein: Philosopher-Scientist*, ed. Paul Arthur Schilpp (Evanston, Ill.: The Library of Living Philosophers, Inc., 1949), p. 176. Born replies with further appeal to God. See also the discussion of "simplicity as a criterion for reality" in Henry Margenau, "Einstein's Conception of Reality," *Albert Einstein*, pp. 255ff.

Einstein is right that God does not throw dice. God has no need of any resources beyond himself in order to decide with perfect wisdom all that he will do. However, this is not a valid objection against quantum mechanics, since the uncertainty built into the quantum mechanical formalism represents principial uncertainty in *man*'s knowledge of what God will do, not an uncertainty on God's part. Here is one more instance of the effect of the doctrine of God on physics.

14. Cf., however, other possible reconciliations of biblical miracles with present-day observed regularities—Bernard Ramm, *The Christian View of Science and Scripture* (Grand Rapids: Eerdmans, 1954), pp. 156ff., 229ff.

15. For a fuller discussion on Contrast, Variation, and Distribution of linguistic Units smaller than the sentence, see Pike, *Language*, especially pp. 154ff.

16. Contrast this with Dooyeweerd's distinction of two kinds of thinking (see Appendix 2).

Chapter 6

STUDY AND ITS ETHICS

We are now ready to reflect on the way in which scientific activity fits into the program of the kingdom of God that God has committed to his people in this Corporate Developmental Application Period. We will approach the question using Particle, Wave, and Field Views, leading successively to modal (6.1), temporal (6.2), and structural (6.3) discussions of science.

6.1 Particle: description of science in relation to forms of Personal activity

I will discuss the problem of the flexibility and fluidity of terms like 'science' (6.11); a proposed classification of science, philosophy, and theology (6.12); and the relation of science, philosophy, and theology to one another (6.13).

6.11 The problem of terminology

Among the debated questions in Christian philosophy of science are (a) what is the nature of science, (b) what is the purpose of science, and (c) what is the relation among special sciences, philosophy, and theology.[1] These questions have no *unique* answer, because of vagueness and difference in usage in the terms 'science,' 'philosophy,' and 'theology.' The term 'science,' for example, could be used either to include or to exclude social sciences or humanities; it could include or exclude pseudosciences like astrology, alchemy, and witchcraft; it could include or exclude unbelieving "scientists" (because their activity is not for God's service); it could include or exclude less Refined forms of knowledge such as the "sciences" of non-Westernized cultures. Similarly, 'theology' can mean Theology Proper (see 2.1) or science of religion[2] or study of what the Bible says or Sapiential study. In everyday language these terms do not have a completely precise meaning, and their very vagueness suits the purposes that people have of talking about vaguely sensed differences between broad, overlapping areas.

Moreover, a person could choose to define science, or philosophy, or theology in a special, more precise sense, for the purpose of technical discussion.[3] There would be nothing wrong with such a definition, provided that thereafter the person was always careful to use the term in the same sense, or to indicate when he relapsed into the ordinary meaning or some combination of the ordinary meaning with his special meaning. He might even define a term in a deliberately "non-Christian" way without there being anything wrong in what he did. For example, he might say, "In the subsequent discussion, I will use 'reason' in the sense of Kant," or "I will use 'God' in the sense of Aristotle." Then he might proceed to discuss Kant or Aristotle from a Christian point of view.

Hence, for my purposes, it is no use carrying on a learned debate about what 'science,' 'philosophy,' and 'theology,' "ought to" mean. However, it *is* useful to see more closely how science, philosophy and theology fit into a biblical picture of man's task. To this end, I feel it useful to suggest some technical terms of my own that come nearer to describing the type of activity and works that people usually have in mind in using the terms 'science,' 'philosophy,' and 'theology.' At the same time I wish to indicate how such activities can be integrated into a more balanced biblical perspective than people usually have.

6.12 *Some classification of Study*

First of all, science, philosophy, and theology are Prophetically Weighted activities, at least more than Kingly or Priestly. Hence I begin with a threefold distinction.

> Description. *Study* is Personal activity with Prophetic Weight, or the result of such activity. Similarly, *Technics* and *Beneficence* are Personal activities with Kingly and Priestly Weights, respectively, or the results of such activities.

I include the "result of such activity" so that, for example, books are Studies, machines are Technics, and art objects are Beneficences. At least in part, this is in conformity with ordinary usage, where a book may be titled "A Study of Plants," a machine may be titled "a thresher," and an art object may be titled "An Appreciation of a Wooded Scene." Moreover, 'science,' 'philosophy,' and 'theology' are terms used both for the activities of investigators ("he's doing philosophy") or the result ("a science book," "he's reading philosophy"). Note that Technics differs from Techn*ology*, which is the study of one Function with which Technics is Weighted. Moreover, "Study" includes both study and communicating of what is studied, as in teaching. Thus the term 'Study' covers a broader area than the term 'study' (the ordinary English word).

> Description. *Refined* Study, *Refined* Technics, and *Refined* Beneficence consist of Refined parts of these activities and results. This is where special attention is paid to generality of scope, method of engaging in the activity and of obtaining results, and justification for the activity and results.

Refined Study is exemplified by science, Refined Technics by the modern factory, and Refined Beneficence by professional art, charity, and monetary control. However, the terms are intended to cover a broader range than just these exemplications.

The descriptions above have not distinguished between Study by covenant-keepers and by covenant-breakers. But the distinction between these two is all-important, as we saw in 5.33. If we desire to make this distinction explicit, we can do it as follows:

> Description. *Genuine* Study, or Technics, or Beneficence, is such Study, Technics, or Beneficence *for* God and his kingdom; *Pseudo* Study, Technics, or Beneficence is against God and his kingdom.

Note that Study cannot always be simply pigeonholed as *either* Genuine *or* Pseudo. The unbelieving scientist's activity is Pseudo Study, but the result (in terms of articles, books, and knowledge) may be largely Genuine, because he does not succeed in escaping God and the knowledge of God. This accounts in part for the difficulty that covenant-keepers frequently have in winnowing truth from error in the works of covenant-breaking scientists. The same applies, *mutatis mutandis,* to art or Beneficence by covenant-breakers.

6.121 *Science*

Now we are ready to talk more specifically about science.

Description. *Modal* Study is Study whose subject-matter is (chiefly or focally) modality. Similarly, one may define Ontological Study, Temporal Study, Structural Study, and Axiological Study. These overlap.

Description. *Natural Science* is Refined Modal Study by Cosmic Men, of Behavioral, Biotic, and Physical Modes.

In other words, Natural Science is Refined Behaviorology, Biology, and Physics by Cosmic Men.[4] I think that this description fairly well correlates with what people ordinarily mean by "natural science."

Description. *Social Science* is Refined Modal Study by Cosmic Men, of the Personal Mode and various Functions within it, especially when such Study has methodological similarity to Natural Science. When such methodological similarity is at a minimum, we speak of *Humanities.*

By characterizing Social Science and Humanities as Modal Studies, I do not desire to deny their strong interest in structural and sometimes temporal elements.

As an example of temporal interest, take history.

Description. *History* is Refined Study by Cosmic Men, of the (temporal) past of the Human Kingdom, particularly the Technical past.

Most of the time, 'history' is used in approximately this sense. However, people also speak of (e.g.) the "history" of the solar system, showing that 'history' can be broadened to mean practically "Study of the past."

Description. *Science* is Natural Science and Social Science.

By saying that Science is a form of *Refined* Study, we do some justice to the difference between a college physics or sociology course on the one hand and Physical or Social instruction to a child on the other. But we draw no *sharp* boundary between the two.

6.122 *Philosophy*

Now we are ready to attempt to describe philosophy. It has often been noted that what people typically describe as "philosophy" is occupied with many questions of a very general or abstract character, and furthermore that philosophy seems never really to put these questions to rest with a definitive answer acceptable to all.[5] Now, these broad characteristics can actually be used to delineate in rough fashion what philosophy is about. Thus:

> Description. *Special* Study is Study focused on some agreed upon subject-matter with some agreed upon methodology and justification. The agreement takes place within a group of Students that may be large or small. If the agreement is more or less explicit or conscious or well worked out, we have a case of Special Refined Study.
>
> Description. *Boundary* Study is Study that concentrates on questions of a "boundary" character, that is, questions that, in a given temporal stage of history, cannot reach definitive resolution by Cosmic Men.

Boundary Study deals with questions concerning which Students have not yet reached essential agreement even about the *method* of resolution. To use Thomas Kuhn's terminology,[6] they are questions bound up with a choice between paradigms or disciplinary matrices, rather than questions (of Special Study) answered on the basis of existing paradigms. Hence the domain of Boundary Study may vary from age to age and from group of Students to group of Students. For example, questions of Physics that were Boundary for Democritus are Special Study questions for modern physicists of the West. The historicity of the resurrection of Christ is Special Study from the standpoint of the community of Christian (Genuine) Students, but Boundary Study from the standpoint of the whole Western culture.

> Description. *Philosophy* is Refined Boundary Study by Cosmic Men.

Similarly, Technics might be distinguished into Boundary Technics ("Exploration") and Special Technics.

6.123 Theology

To describe theology is somewhat more difficult than to describe science or philosophy. The easiest place to start is with the teaching function of the church. Theologians are the Refined teachers of the church, communicating (or attempting to communicate) the teaching of the Bible. Teaching the Bible has a special role with respect to all other teaching, because in the Great Commission Jesus mentions "teaching them to observe all that I have commanded you," and "all that I have commanded you" is included in the Bible.[7]

In terms of the Dominical, Covenantal, and Servient Views (3.332), we can distinguish three forms of Study by Men. We have already mentioned the Dominical, Covenantal, and Servient Word of God (3.3243). The Dominical Word of God is everything that God says, the Covenantal Word of God is everything that God says that he shares with man in Covenantal speech to man, and a Servient Word of God might be what Jesus Christ says in his Cosmic life, or else what God's inspired servants the prophets say with divine authority, or else Bible teachers' (fallible) attempts to communicate God's Word.

Now the Bible, the Covenantal Word of God during the Corporate Developmental Application Period, is Study. We may distinguish it from other forms of Study by calling it Canonical Study.

> Description. *Canonical* Study is the Covenantal Word of God, or the act of its production.
> Description. *Speculative* Study is the Study of the Dominical Word of God (especially as this goes beyond the Covenantal Word of God), with the purpose of communicating what this Word says.
> Description. *Evangelical* Study is Study of the Covenantal Word of God, with the purpose of communicating what this Word says.

The purpose clause is included in the latter two definitions, so as to exclude (say) from Evangelical Study counting the number of words or letters in the Bible. Counting the letters can be Study, but not Evangelical Study.

> Description. *Theology* is Refined Evangelical Study by Cosmic Men.

What is the difference between Speculative and Evangelical Study? As usual, the line between them is fuzzy. Some more speculative parts of theology might plausibly be classified as either, depending upon one's viewpoint. What, for example, are we to say about study of the relation of geology to the Genesis 1 account? Certainly the results of such study may be partly Speculative, but they are of interest to the Theologian.

But some cases are clear. For example, the teaching of justification by faith is Evangelical Study. Similarly, the finding or communicating that $F = ma$ is part of Speculative Study. $F = ma$ is not part of the teaching of the Bible (if it were, it would be Canonical Study). It is a "speculative" approximate, tentative description of God's decree (Dominical Word of God) for the dynamics of moving objects.

"Theology" includes not only systematic theology but also the sister disciplines of biblical introduction, exegesis, and biblical theology.

Description. *Systematic Theology* is Theology that answers the question, "What does the Bible as a whole say?"
Description. *Exegesis* is Evangelical Study that answers the question, "What does this [some particular] passage say?"

The above descriptions of Science, Philosophy, and Theology are supposed to conform fairly closely to the ordinary meanings of 'science,' 'philosophy,' and 'theology' in English. By contrast, this is *not* what most of the other technical terms (capitalized terms) are necessarily supposed to do. Sometimes, indeed, a correspondence is fairly obvious. For example, "biology" in English would be something like "Refined Biology" or perhaps "Special Refined Biology." On the other hand, "logic" in English is something like "that part of Refined Logic that studies the formal structure of arguments"; "physics" is something like "the part of Special Refined Energetics that is not chemistry"—though in physics considerable attention is paid to Kinematics and Mathematics. Hence the terms 'Logic' and 'Physics' do not match ordinary English terms.

6.13 *Interlocking of Science, Philosophy, and Theology*

It should be clear even from the descriptions themselves that Science, Philosophy, and Theology are intimately related to and overlap with one another. We could, in fact, speak in more general terms of the overlapping and interrelatedness of Modal Study (of which Science is a part), Boundary Study (of which Philosophy is a part), and Evangelical Study (of which Theology is a part). But our examples might as well be from Science, Philosophy, and Theology.

6.131 Modal Boundary Studies

Every Science and every Modal Study contains Boundary problems. In Physics, what is the nature of mass? of the ultimate material constituents of Inorganic Creatures (or are there such things?)? Why and how does Mathematics apply to Kinematics and Energetics? In Linguistics, what is the nature of symbols? How can and do symbols acquire meaning? What is meaning? And so on.

6.132 Modal Evangelical Studies

What the Bible says about Biology ought to be reckoned with in Evangelical Biology, what the Bible says about language ought to be reckoned with in Evangelical Linguistics, and so on. But Modal Speculative Studies go beyond what the Bible says about their subject-matter.

6.133 Evangelical Boundary Studies

Some Evangelical Study is not really Boundary Study. A study, for example, of the prosaic matters in the reigns of the kings of Judah is Special Study. Moreover, almost no Evangelical Study is of Boundary character if one limits consideration to a group of Students committed to the authority of the Bible. (In this case, they can agree on methodology.) However, in terms of the West as a whole, Evangelical Study of God, Christ, the resurrection, etc., is Boundary Evangelical Study.

6.134 Communication among Science, Philosophy, and Theology

It is difficult to generalize about the mutual interplay of Science, Philosophy, and Theology. We cannot easily predict what light an insight in one of these fields might throw on problems in another field. In any case, communication and sharing among Genuine Studies can only result in mutual benefit. However, the introduction of Pseudo Study causes difficulty. Someone ill acquainted with a Special Study can only with great difficulty recognize elements of Pseudo Study in this Special Study. The problem is difficult enough *even* for one *well* acquainted with the Special Study. Thus it is understandable—and to some extent necessary—that suspicions develop between different fields, and that one field of Study approaches the "assured conclusions" of another with skepticism.

Evangelical Study has a special role, as over against Speculative Study. Speculative Study must, well, speculate. Since it does not limit itself in *sola scriptura* fashion to the Bible's teaching, it cannot press home its claims with the same pattern of appeal to divine authority. Evangelical Study starts with a linguistic corpus with divine authorship and authority, and must simply translate and paraphrase this corpus to help people see how the Bible applies to their situation. Speculative Study, or at least Genuine Speculative Study, uses the Bible indeed, but must also (typically) work from some nonlinguistic effect of God's Word *toward* a human verbal formulation. For example, a physicist starts with motions of Creatures, which motions God has ordained. Then he comes to a linguistic formulation like "F = ma."

Nevertheless, the boundary between Evangelical and Speculative Study is fuzzy. An originally Speculative statement, for example, may become Evangelical during the course of church history, because Theology has grown more sensitive to what the Bible teaches. For instance, an initially Speculative statement—not claiming full scriptural support—about mode of baptism or the Antichrist may later be seen to be supported or refuted by Scripture.

Finally, the extra deference given to Evangelical Study is still not absolute. It is true that, if we reach the conclusion that the Bible teaches something, we have the obligation to live on the basis of that teaching. However, this is *not* the same as saying that Speculative Study cannot be useful both (a) in refining our understanding of the Bible (by Study of biblical languages and archaeology) and (b) in leading to a re-examination of whether the Bible *really* supports a view that Speculative Study appears to be pointing away from.

For example, a person who believed that the Bible taught that the earth had four literal geometrical corners could be led by Speculative Study of Geography to re-examine the Bible's teaching. Then, especially if he consulted those more gifted in their knowledge of the Bible than he, he would realize that the Bible's language at this point is metaphorical. So an adjustment would have taken place in his Evangelical knowledge on the basis of Speculative Study.

Take another simple example. It has been Speculatively claimed that the patriarchal narratives of Genesis make an anachronistic error in saying that camels were ridden at that time. What should be a believer's response? The believer first of all might recognize that in this case the Speculative evidence is not nearly so strong as it is about the "four corners." However, he may be led to re-examination of the Bible, with the questions: (a) are the patriarchal stories mentioning camels presented by the Bible as fictional or historical? (b) has the time and place of the patriarchal events been correctly reckoned? (c) does the word 'camel' in the original really always mean "camel" in the modern sense? But in this case, re-examination leads to the conclusion that the Bible *does* say that they rode camels. Hence the Speculative claim to the contrary is dismissed as Pseudo History. The one thing that ought never to be done is to say, "The Bible is mistaken."

In a more difficult case, of course, a believer may have to say, "I don't know how to reconcile the Bible and Speculative Study." In that case, because Speculative Study is contaminated by Pseudo Study, he ought to trust the Bible and, until more light can be thrown on the question, disbelieve enough of the less plausible parts of the Speculative Study to bring it into conformity with what he is sure the Bible says. If, on the other hand, he is unsure of what the Bible says on a particular point, he may have to remain in suspense.

All this may not be satisfactory to those who desire a quick and easy answer to the science-religion conflict, and who want a way of putting down the skeptics. But (a) Scripture admonishes us to be content (Phil. 4:11ff.; Eph. 5:20). We ought to be content with the fact that God is more wise than we are about just what answers will be for our good (Rom. 8:28). He is fully able to save skeptics even without answering all covenant-breaking scientists. (b) If one of us *were* able to answer all the challenges from Pseudo Science, it would not save anyone (Luke 16:31). (c) All doubts will be put down in the consummation. Now is a time when we suffer with Christ (Phil. 3:10; II Cor. 1:5), and part of this suffering is the trial of our faith (I Pet. 1:7). It is God's purpose that faith should be tried by means of situations where the temptation to doubts is present.

6.2 *Wave: historical development of Study*

A deeper appreciation for the Ethics of Study can be gained by attention to the examples and teachings in the Bible about Genuine and Pseudo Study. We have already discussed this to some extent in 3.3432. Now I will draw out some Ethical implications of 3.3432. Let us consider some of the Periods one by one.

6.21 *Study in the Adamic Accomplishment Period*

The first Student is God, who calls his Creatures into being (Rom. 4:17). One of the general principles of Scripture is that man's work is to be an imitation of God's (Eph. 5:1). His Study ought to be holy as God's Study is holy (I Pet. 1:15-16).

The first recorded instance of Man's Study is the Passive Study of listening to God's commands (Gen. 1:28-30). Next, Adam engages in Active Study of the taxonomy of the Animal Kingdom. The order of these two activities may be important. The inspired prophet in Scripture must hear the Word of God before he speaks it (Exod. 7:1-2; Num. 12:8; 23:5, 8, 12, 27; 24:4; Deut. 18:15, 18; John 8:38; 12:49-50; Eph. 3:3-5). Likewise the Prophet in the broad sense must know God in order to be able to imitate him.

Adam's naming involves an element of Technics and Beneficence as well as Study. How so? Adam's naming, like God's (Gen. 1:5, 10), is an exercise of sovereignty, thus fulfilling Genesis 1:28. Now God's word both characterizes what a thing is like (its meaning; Locutionary), and defines what function it shall play in Creation (its power; Administrative), and appraises its value (Sanctional). Hence we expect that Adam's naming will also involve these three elements, though naturally the Locutionary element will be most prominent because "naming" is Weighted in the Prophetic Function. The one instance of naming in Genesis 2:23 confirms this impression. Woman is not only distinguished from the animals, but is also described in terms of her function or role *vis-a-vis* Adam. And the delight, the appreciation, that Adam has for his wife is not hard to detect in his words.

Thus, in the beginning, both in the directives that God gives to Man and (so far as we can tell) in Man's fulfillment of them, there is an integration of Study, Technics, and Beneficence. If we were to put it in modern terms, we would say that there is an integration of science, technology, and art—but this would seriously narrow the meaning. Remember that Study, Technics, and Beneficence, as this book uses the terms, are broader than modern science, technology, and art.

Now, what justification can there be for a purely "theoretical" Refined Study or Science that would be devoted to achieving knowledge purely for the sake of knowledge? None. All Genuine Refined Study aims at "filling the earth and subduing it, and having dominion over the fish of the sea, and over the birds of the air. . . ." In Refined Study each man should be aiming both toward extending his own dominion and toward equipping others with the knowledge and skills necessary for exercising dominion. Adam is not told to hope for some absolute, objectively perfect theoretical world-system as the end of Refined Study; rather he is working for a kind of application of truths about the Cosmos to the needs arising in connection with ruling.

Therefore, Adam's Study was in crucial respects *superior* to ours—in spite of all the inflated boasts about modern science. Certainly it was superior in its conformity to the standards and goals of God, superior also perhaps in its integration of "theoretical" and "applied" science. It is not true that ancient man was fatally condemned to muddle through as best he could in ignorance and superstition, until the rise of modern science. It was not true of Adam, nor was it true of ancient Israel, who were entrusted with God's Covenantal Words including those words to Adam.

6.22 *Study in the Adamic Application Period: the fall and Study*

Since fallen man wishes to be his own god, he does not want to obey the mandate of Genesis 1:28-30 in its true sense. In Study, his rebellion can take the form of (1) refusing to Study (laziness; an ontological rebellion), (2) substituting a Pseudo Study (superstition; an axiological rebellion), or (3) using Study for evil ends (perversion; methodological rebellion). These three correspond to similar options in rebelling against God's written word by (1) ignoring it, (2) substitution a Pseudo word of God (e.g., the Koran or the Book of Mormon), or (3) using the word for evil ends (hypocrisy). These three responses, of course, are interrelated and mutually complementary (cf. 5.132).

God indicates that in spite of rebellion, man will continue, yes, must continue to fulfill Genesis 1:28-30 in some sense. He will be fruitful and multiply (Gen. 3:16), and subdue the earth (Gen. 3:17-19). This becomes even more explicit in the promise made to Noah (Gen. 9:1-7).

6.23 *Study and Technics in the Preparation Period*

I have already mentioned the chief examples of Study in the Preparation Period (cf. 3.3432): Noah, Joseph, Moses, Solomon, Elijah, Isaiah, etc. There are also examples of Technics, sometimes integrated with Study: Jabal, Jubal, and Tubal-cain (Gen. 4:20-22), the Nephilim (Gen. 6:4), Noah (6:14-22; 9:20-21), Nimrod (10:8-9), and Babel (11:1-9).

Some of the Study is even Refined. Moses (especially in Num. 33) and the authors of the OT historical books provide examples of History (note the use of sources in Josh. 10:13; II Sam. 1:18; I Kings 11:41; 14:19, 29; 15:7, 23, 31; 16:5, etc; I Chron. 9:1, etc.). Psalms 78, 105, and 106 are also devoted to History, probably with the Pentateuch as a source. Portions of Psalms 104, 147, and 148 are devoted to Natural Science (though one might argue that it is not very Refined). Solomon appears to have been the greatest Natural Scientist of them all, according to I Kings 4:29-34.[8] Proverbs and Ecclesiastes represent Social Science (note the methodological care spoken of in Eccles. 12:9-11).

6.24 *Study in the Accomplishment Period*

Christ is the final and perfect Student. This is already implied by the fact that he is the final Prophet, as we have seen (3.3432). His Study, that is, his teaching, is Refined. First, it has generality of scope (John 15:3, 15; Luke 24:44, 45, 47; Col. 2:3); it covers all of a believer's life. Second, he is self-conscious and reflective about his methodology in communicating the truth. For instance, Christ specifies the source of his teaching (John 12:47-50), why it takes the form that it does (Matt. 13:11ff.), and what is its goal (John 6:63). Finally, he gives grounds of justification for his language in the commission of his Father (John 12:49-50; 5:30-47; 7:16-18; 8:14ff.).

The claim that Christ is the archetypical, the representative Student for the world, may seem strange from the standpoint of modern science. But if it does seem strange, it is because science has lost its bearings and lost sight of its true purpose in a way that Christ alone has not. He, and not the modern scientist, is the ultimate fulfillment of God's command in Genesis 1:28-30; for he is the seed of the woman (Gal. 3:16), to whom all things are now subject (Eph. 1:20-21; Heb. 2:6ff.; Matt. 28:18-20).

6.25 *Study in the Application Period*

In the Generational Application Period, Luke is the outstanding Historian, and Paul the outstanding Theologian-interpreter of the OT. Luke's Study is Refined, as can be seen from Luke's attention in Luke 1:1-4 to generality of scope ("all things closely for some time past"), to methodology ("orderly account"), and to justification for what he says ("eye witnesses"; "that you may know the truth"). Similarly, Paul shows Refinement especially in the Book of Romans with its methodological arrangement and close argumentation.

I have said before that the Bible as a whole is Canonical Study. But in fact it is *Refined* Canonical Study. It is truth put into a form suitable for communication to the whole Cosmic Human Kingdom. (a) Ontologically, it has generality of scope: it is intended for the church of all ages, it contains many general unqualified truths, and it contains a thorough discussion of who God is and of the development and formation of the people of God in history. (b) Methodologically, it gives guidelines for the preservation of the Word of God (Deut. 4:2; 12:32; 31:9ff., etc.) and the recognition and reception of true and Pseudo prophets (Deut. 13:1-8; 18:9-22; John 5:30-47; 10: 31-39, etc.). (c) Axiologically, it gives ground for its statements in the inspiration of the prophets and the faithfulness of God.

Furthermore, much of Scripture is of Philosophic character: for instance, its discussion of who God is, of God's moral demands, of the meaning of history, of the meaning of the death and resurrection of Christ, and of the second coming of Christ. We would not know these things (or at least we would be very much in the dark about them because of sin) apart from the Bible. Those in the West who refuse to take the Bible for their guide dispute all these things. But it still may seem odd to call the Bible "Canonical Philosophy." Why? The answers that Pseudo philosophers have proposed, and even the methods and justification connected with their proposals, are so radically different from the Bible. But does not this only show how far people can stray from God? Sometimes a sinner does not even know what questions to ask, much less how to get a reliable answer.

Some people may object that my comparison between the Bible and modern philosophy and science does not show appreciation for obvious differences. First, the Bible has a marked redemptive-historical thrust. It pays special attention to God's Covenantal Locutions, Administration, and Sanctions in their historical unfolding. Second, from beginning to end the Bible bears witness to Christ.

My reply is as follows. First, the Canonical status of the Bible indeed makes it special. The OT types and prophecies point forward to Christ in a way that could not have taken place apart from divine authorship. The OT and the NT are the Words of God as well as the word of men. This I grant. But why should we assume that modern science or philosophy cannot have a redemptive-historical thrust, and cannot bear witness to Christ? Precisely this assumption I challenge.

In fact, I would go so far as to say that any so-called Science that does not bear witness to Christ is sheer Pseudo Science. Modern Science or Refined Study ought to bear witness to Christ, albeit in a way derivative from Scripture. Naturally, it does not compete with Scripture by trying to do what Scripture itself does. It Speculatively applies Scripture to the situations and questions at hand. And Genuine Refined Study also has a redemptive-historical thrust, derivative from Scripture. It knows that we are living in the days of the Spirit, looking and working for the consummation, and interprets the progress of history ("secular" history) and the course of its own discoveries and failures, in the light of God's purposes expressed in Scripture. Thus modern Refined Study might well end up sounding like Psalm 104 (e.g., it might contain explicit reference to God, to creation, and to the praise of God; it might be written as poetry—so integrating itself with Beneficence).

6.3 Field: Study in relation to the structure of human activity and the kingdom of God

This section is devoted to the question of what God requires of us with respect to choices within Study and choices between Study and other activities. Let us begin with an example.

Suppose that Ralph is a student in college, thinking about career plans. What area of Study does the Lord want Ralph to choose for a lifetime work? Where can he make the most significant and lasting contribution to the kingdom of God? Or suppose that Bob is a biologist working for the Department of Agriculture. Should he give up his job to become a preacher? Should he concentrate on communicating the gospel to his colleagues? How can he change what he is doing on the job to make it "Christian"?

What Christian is there who has not wrestled with questions like these? We begin to answer them biblically when we have some appreciation for how Study and science fit into God's "marching orders" for the church in the Corporate Developmental Application Period. But a good deal will depend upon the particular time (see 6.2), particular circumstances (such as the existing state of Science; see 6.33), and particular talents of men. In different Special Sciences the details of thinking through a Christian viewpoint may be very different. Hence in a book of this size we must confine ourselves to generalities. We are dealing with Ethical questions, so the questions can be viewed from the Existential 6.31), Normative (6.32), and Situational (6.33) Perspectives (recall 4.1).

6.31 *Calling from the Existential Perspective*

Under this head we consider the *persons* involved in Study. What directives does God give to them? Before all else, God invites and commands us to believe in him whom he has sent, his Son Jesus Christ (John 6:29). This means resting on and trusting in him for salvation from sin and the wrath of God. It implies that

> I, with body and soul, both in life and in death, am not my own, but belong to my faithful Saviour Jesus Christ, who with his precious blood has fully satisfied for all my sins, and redeemed me from all the power of the devil; and so preserves me that without the will of my Father in heaven not a hair can fall from my head; yea, that all things must work together for my salvation. Wherefore, by his Holy Spirit, he also assures me of eternal life, and makes me heartily willing and ready henceforth to live unto him.[9]

Thus, if Ralph or Bob is not a believer, believing is the *first* thing he must do. Until he believes in Christ, all his works are futile (John 15:5). And such belief is not simply an assent to what the Bible says, but involves a life-transforming personal relationship to Christ. A believer cannot, will not, remain in his sins (I John 3:6-7), because the Lord has given him a new heart (Ezek. 36:26f.).

Second, if we believe in Christ, we must learn his words and keep them (John 15:7-11). We must pray continually (I Thess. 5:17; Rom. 12:12; Eph. 6:18). We must take up God's armor (Eph. 6:10ff.). And so we could go on. These are the "marching orders" of the church of Jesus Christ, since his ascension. They are summed up in those famous words of our Lord,

> All authority in heaven and on earth has been given to me. Go therefore and make disciples of all nations, baptizing them in the name of the Father and of the Son and of the Holy Spirit, teaching them to observe all that I have commanded you; and lo, I am with you always, to the close of the age [Matt. 28:18-20].

Hence Study should first of all be study and teaching of what Jesus Christ commands.

Moreover, the *way* in which God's "marching orders" will be fulfilled is described in the last words of our Lord on earth,

You shall receive power when the Holy Spirit has come upon you; and you shall be my witnesses in Jerusalem and in all Judea and Samaria and to the end of the earth [Acts 1:8].

Hence this Study is through the power of the Holy Spirit.

What should strike us and thrill us is that the church's task, our task, is Spiritual. That does not mean that it is ethereal, or insubstantial. It does not mean that we are supposed to be so heavenly minded that we are no earthly good. Or that we are not supposed to transform the Cosmos. It means that we need to be empowered by the Holy Spirit, to be engaged in restoring, through the gospel, Man's relation to God through Christ. Our "marching orders" have a Sabbatical accent or Weight, a gospel accent.

That is why a person's service to the Lord does not depend so much on his being in a particular geographical location or in a particular kind of job. No one is ever in a situation where he cannot worship the Father in spirit and truth. The work of slaves is just as precious in God's sight as the work of masters, of children, of fathers, of wives (Eph. 6:5ff.; cf. Col. 3:22ff.). We must no longer evaluate work in the way that covenant-breakers do, in terms of humanism or pragmatism or communism. No. We must ask, "What does God count precious?" And when we study that question, we see that all worldly standards are overthrown (Luke 1:51-55). Whoever receives a child in Christ's name is blessed, and the least is greatest (Luke 9:48). We must be a slave and servant rather than lord it over others (Matt. 20:21-28). Jesus himself sets the example by washing the disciples' feet (John 13:1-20). Jesus counts as precious the service of love and devotion that the world counts as waste (John 12:1-8; Luke 21:1-4).

The church is the people of the kingdom of God. Such service of love is the work of the kingdom. But the world does not recognize this as a kingdom at all, because it comes in weakness and suffering (John 18:33-38). The church itself is in danger of falling into the Corinthian error of conceiving its work in different terms (I Cor. 4:8-13). There is the constant temptation for the church to become a power bloc over against other power blocs, trying to lord it over Men; or for an individual believer to "become rich," to "become kings," to "reign" (I Cor. 4:8).

Of course, the civil magistrate, whether he be a believer or an unbeliever, has the obligation to exercise the power of the sword (Rom. 13:4). But such a "sword" or such an authority has *not* been committed to the church (that is, the NT people of God), or to an individual believer outside of such a definite God-ordained office.

At this point an objection arises. Some people would claim that the "marching orders" of the church include fulfilling the "cultural mandate" of Genesis 1:28-30 and 9:1-7, in addition to the "evangelical mandate" of Matthew 28:18-20. But the words 'in addition to' are not felicitous. Those words can betray a misunderstanding of the bearing of Genesis on our age. Since the commands of Genesis 1:28-30 and 9:1-7 were given in earlier Periods of history, we must be cautious about the way that they apply now.

As we have seen in 3.3432 and 3.3433, these commands have been fulfilled in Christ who fills all things and has subdued all things under his feet. Consequently, Matthew 28:19-20 describes the form that the church's "filling and subduing" takes in the Application Period. Genesis 1:28-30 and Matthew 28:18-20 should not be played off against one another. Moreover, Matthew 28:19-20 is *based on* the fulfilled work of Christ, who says, "All authority in heaven and on earth has been given to me" (28:18). Therefore, if we are to follow his leading, we cannot avoid numbering ourselves among those under the obligation of Matthew 28:19-20.

On the other hand, neither should we be reluctant to see and acknowledge that all the Christian's everyday work (not just a narrower work of spreading the gospel message) is to be sanctified by the indwelling Spirit of Christ, so that he begins again to fulfill Genesis 1:28-30. Because of the words "*all* that I have commanded," Matthew 28:18-20 cannot legitimately be used as an excuse for conformity to the world outside of a narrowly "religious" area.

All this explains why Paul is not anxious over what status a believer has with respect to the world. He says, "Let every one lead the life which the Lord has assigned to him, and in which God has called him. This is my rule in all the churches" (I Cor. 7:17). Why? Paul tells us: "For neither circumcision counts for anything nor uncircumcision, but *keeping the commandments of God*. Every one should remain in the state in which he was called" (7:19-20). We can serve the Lord where we are, slave or free, by keeping the commandments of God.

Hence, we have already obtained part of the answer to the questions above about people's callings. The biologist should remain a biologist. He need not become a preacher. Indeed, to try to become a preacher would be sinful, if he has not the gifts necessary for such work.

But the Apostle Paul's dictum "remain where you are" must be qualified by what he says here and elsewhere about "keeping the commandments of God." A person who becomes a believer cannot remain in an occupation in which he cannot avoid sinning. For example, "let the thief no longer steal, but rather let him labor, doing honest work with his hands, so that he may be able to give to those in need" (Eph. 4:28).

This also puts a twist even on the situation of a person in a "lawful calling." Suppose that our biologist does have or develop gifts for preaching or teaching or ruling the church. The Bible says that out of love he must exercise these gifts (I Pet. 4:10-11). This is part of "keeping the commandments of God." When this requires a large amount of his time, the church should pay him (I Tim. 5:17-18; I Cor. 9:3-14). Thus it may happen that he becomes only a "part-time" biologist, as Paul was a part-time tent maker (Acts 18:3ff.). Or he may give up completely earning a living as a biologist (I Cor. 9:14).

Thus it is that, when a choice comes between biology and preaching the gospel, preaching must be preferred on the grounds that the central thrust of the Great Commission is "teaching all that I have commanded you" (Covenantal Word of God), not "teaching biology." Study in the NT age is to have an Evangelical center.

Work in science or other fields must also be evaluated in terms of the scientist's or worker's motives. Does he work out of love in the sense of I Corinthians 13? Does he exhibit the fruits of the Spirit of

Galatians 5:22-23? Because of the deceitfulness of our hearts, it is easy for a believing scientists to assure himself that he is doing Genuine Science, that he has not compromised with unbelief, when in fact he has.

Hence the obligations for us to pray, to grow in personal communion with God, and to seek and build a solid fellowship in the church are not as remote from scientific concerns as it might be supposed (James 1:5).

6.32 Calling from the Normative Perspective

Next, let us ask: what commandments does God give concerning Study itself. We have already answered to some extent in sections 6.2 and 6.31. However, what is important to notice is that God's commands give us liberty (cf. Gal. 4:31–5:1, 13). They do not require us to define "science," "biology," or "history" in a certain way. They do not require us to use only the "accepted" scientific method if an alteration should prove useful. They do not require us to value scientific claims and achievements according to an artificial or customary set of values put forward by the world. In other words, the Word of God sets us at liberty with respect to questions of ontology (what is the subject-matter of a Study?), methodology (how do we go about Studying it?), and axiology (what is its value?).

The Bible sets us at liberty not only by providing satisfying ultimate answers, but by remaining silent. The person who sees what *does* matter (6.31) is free to adopt any of a variety of perspectives about issues that do not matter. He may choose whichever alternative he wishes when the Bible does not require him to choose one. For example, he is free to adopt any of the Views or Perspectives in this book, and to use any form of Emphasizing Reductionism that may seem promising. 6.2 makes it clear that in the past Study has occurred in varying ways—many of which have the Bible's approval.

Pseudo Study, by contrast, is often bound to a single viewpoint; it is caught in a form of Exclusive Reductionism. In many cases a human dictum "you *ought* to do it this way" is regarded as binding even though it has no basis in Scripture. For example, consider the field of psychology. The behaviorist argues for a behavioral approach to psychology, the Freudian for a Freudian approach, and the Rogerian for a Rogerian approach. Each says, "Psychology ought to be done this way." Where does the "ought" come from? It should be clear that in the case of psychology (and indeed in the case of a good many scientific disputes), the response of Genuine Science is not simply to say that all three (or more) sides are right. Exclusive Reductionism in its Exclusiveness always distorts in the attempt to claim ultimacy. But Genuine Science, freed from the compulsion to be "ultimate," or to be wise and important in the world's eyes, is able to do greater justice to the Contrast, Variation, and Distribution of subject-matter. It does not reject unpalatable facts because they will not fit over-simplified theories.

But it is not always simple to free oneself from the influence of Pseudo Study. In the last several hundred years, science and philosophy have been dominated mostly by unbelievers and by inconsistent believers. Hence, it is hard for people today not to get caught in those subtle patterns of thinking that in practice deny God or ignore him. We think of "matter" instead of Inorganic Creatures, "nature" instead of God, "art for art's sake" instead of Beneficence. Hence I have written this book in the hope of helping people engaged in Study. Partly by changing vocabulary, mostly by appreciating the biblical teaching on science, one can begin to integrate Study with a biblical world view instead of doing it "autonomously."

Why should we desire this? The Lord Jesus Christ, as *Lord,* claims every bit of science, philosophy, or other Study to himself. In him "are hid all the treasures of wisdom and knowledge" (Col. 2:3; see also 3.3432). Hence Paul says, "We destroy arguments and every proud obstacle to the knowledge of God, and take every thought captive to obey Christ" (II Cor. 10:5). Here is a call to reform all of Study, and to lay it at the feet of Christ our Savior.

In particular, I would stress the importance of what I have called Sapiential Study, that is, Study of matters with particular focus on how God is involved in them. Sapiential Study has in the past suffered both neglect and perversion by Pseudo Philosophy. It is a wide-open field for believing Students to explore.

Who is equal to such a calling? We are unworthy of it. But Paul says,

> Not that I have already obtained this or am already perfect; but I press on to make it my own, because Christ Jesus has made me his own. Brethren, I do not consider that I have made it my own; but one thing I do, forgetting what lies behind and straining forward to what lies ahead, I press on toward the goal for the prize of the upward call of God in Christ Jesus [Phil. 3:12-14].

6.33 *Calling from the Situational Perspective*

Next, what are we to say about a young person who has a choice of vocations still open before him? Or the scientist who must make a choice between two areas of study or two methods? In many situations *either* choice is right (see 6.32). In other situations, one choice may be preferable to another. I Corinthians 7:25-40 illustrates the principle involved in making such decisions.[10] Paul's general principle is, in cases of choice, to choose that station in life most suitable for undistracted worship and service to the Lord. For example, if sexual desires or temptations distract the unmarried person, he should marry. If a person will be distracted by cares for his wife, he should remain single.

Though Paul here deals with a special case, it is obvious that his principle could be applied to a choice of calling by a young person, or the choice of methods by a scientist. The young person should evaluate possible callings in terms of his gifts, his desires, his temptations, and choose that calling in which he may best serve the Lord. This is the same as choosing the calling in which, as one member of the body of Christ, he may help in fulfilling Matthew 28:19-20. This same criterion, naturally, is also to be used in cases of less momentous decisions within a program of Study.

This may sound like advice for everyone to become preachers, but it is not. A further reflection on the Bible's teaching will show that it is very far from being that. For one thing, not all have the same gifts (I Cor. 12:4-31). Every member of the body is important, valuable, yes, indispensable. No one should think himself inferior or worthless because his particular contribution to the body is less prominent.

Indeed, those with prominent gifts have greater responsibility (Luke 12:48) and greater temptation to pride (I Cor. 12:21).

But a person will also have to be careful about what he calls a spiritual gift, and what he judges as useful to the body of Christ. A gift, for example, in understanding biology or political science is not by itself a "spiritual gift." Unbelievers may also have such gifts. They are gifts from God, but that does not necessarily mean that they are indispensable to the healthy functioning of the body of Christ. No talent that a believer shares with unbelievers is a spiritual gift in the sense that Paul is talking about in I Corinthians 12.

Nevertheless, there are still reasons why a believer might choose to Study (say) biology. For one thing, a believer must take into consideration his degree of interest and zeal in different possible callings. The Apostle Paul commands us to work for our living (II Thess. 3:12; I Thess. 4:11-12). In such work we should work with all our heart (Col. 3:23f.). All this would be more difficult in a work that we are not interested in or less gifted in.[11]

Secondly, in connection with almost any vocation a believer can exercise some of his spiritual gifts. For example, he may through his Study of Biology be able to exercise Christian compassion in the medical profession. Or by teaching Biology he may help others to do so. Because of his submission to God and the Bible, he may be able to root out some Pseudo Biology that formerly created a stumbling block by opposing the Bible's teaching.

Moreover, as a teacher he may have a special gift for appreciating and being able to communicate to others the wonder of God's wisdom in making the human body or various kinds of animals and plants. To be sure, his speaking about God's wisdom in the Creation is not the central element in the gospel or in the Great Commission. But it does help to place the gospel in the right setting by explaining who is the God of whom the gospel speaks. And it may help believers more effectively to obey the injunction to be always thankful and to praise God in their prayers. A larger vision of God's greatness and sovereignty can inspire our hearts, embolden our witness, and enhance our zeal in ways that we cannot always anticipate.

NOTES TO CHAPTER 6

1. See, for example, Abraham Kuyper, *Principles of Sacred Theology* (Grand Rapids: Eerdmans, 1968), especially pp. 59-105, 228-340; and Herman Dooyeweerd, *In the Twilight of Western Thought* (Nutley, N. J.: The Craig Press (1965), pp. 113-172. Unfortunately, both of these works are seriously damaged by the tendency to believe that the words 'theology' and 'science' have one correct sense. In particular, Kuyper appears not to have grasped fully the fact that etymology is not a norm for present meaning and synchronic analysis.

2. When Dooyeweerd defines "theology" as the science that studies the pistical aspect—*A New Critique of Theoretical Thought* (Philadelphia Presbyterian and Reformed, 1969), II, p. 562—he appears to make "theology" mean what the world in general calls "science of religion." Of course, Dooyeweerd has a special meaning for 'religion,' so that he himself would not use the phrase 'science of religion.' For discussion of the effect of this terminology in cosmonomic thought, see the discussion of 'theology' in 9.2.

3. Cf. the discussion of the implications of introducing technical terms in Carl G. Hempel, *Fundamentals of Concept Formation in Empirical Science,* International Encyclopedia of Unified Science, vol. II, no. 7 (Chicago: University of Chicago, 1952).

4. Technically speaking, to make the descriptions match one would have to redefine Behaviorology, Biology, and Physics as *Study* (instead of *study*) of the Behavioral, Biotic, and Physical Modes respectively. But the reader can easily supply these changes where necessary.

5. See especially Hendrik van Riessen, *Wijsbegeerte* (Kampen: Kok, 1970), p. 11.

6. Thomas S. Kuhn, *The Structure of Scientific Revolutions,* 2nd ed. (Chicago: University of Chicago, 1970), especially pp. 182-187.

7. Cf. the note on the closure of the canon in 3.32432 and the discussion of the sufficiency of Covenantal Ethics in 4.2.

8. Onomastica, or lists of names, were a customary form of wisdom about the "cosmic order" in the ancient Near East. Cf. Georg Fohrer, "Σοφία," *Theological Dictionary of the New Testament,* VII, ed. Gerhard Friedrich (Grand Rapids: Eerdmans, 1971), p. 479.

9. The Heidelberg Catechism, Question 1, from Philip Schaff, *The Creeds of Christendom, with a History and Critical Notes,* III, reprint (Grand Rapids Baker, 1966), pp. 307f.

10. The qualifying statements in I Cor. 7:12, 25 do not, in my opinion, indicate that this portion of I Cor. is not inspired, but rather that Paul is giving his inspired judgment and counsel (7:40b!) on issues about which Jesus Christ the Mediator did not specifically speak while he was in the Cosmos. Cf. the discussions in John Murray, *Principles of Conduct; Aspects of Biblical Ethics* (Grand Rapids: Eerdmans, 1957), pp. 68ff.; Charles Hodge, *A Commentary on the First Epistle to the Corinthians,* reprint (London: Banner of Truth Trust, 1964), pp. 114, 126; Archibald Robertson and Alfred Plummer, *A Criti-*

cal and Exegetical Commentary on the First Epistle of St Paul to the Corinthians (Edinburgh: T. & T. Clark, 1911), p. 141; John Calvin, *Commentary on the Epistles of Paul the Apostle to the Corinthians* (Edinburgh: Calvin Translation Society, 1848), pp. 240f., 252f.

11. But asking the Lord to give us more interest in and zeal for a chosen occupation is always a possible alternative to changing occupations.

Chapter 7

CONCLUSION

In conclusion it is well to reflect on where we have been (what does this book as a whole say?) and where we are going (what is next?). I discuss the second question first.

7.1 *The next steps*

When I began this book, I intended to write about philosophy of mathematics. But it soon became apparent that many Philosophical questions had to be dealt with, in order for me to give an adequate explanation for why I would say and do what I do with respect to mathematics. And so I have discussed mathematics scarcely at all. But what I have said here *can* serve as a preliminary to many Special Studies. Indeed, it would be most useful to illustrate how the Bible's teaching affects various Special Studies. Not only that, but some Special Studies can help to reinforce what I have tried to say here.

I would single out several areas as particularly important objects for reform: apologetics, linguistics, logic, and probability theory. Apologetics, because avoiding compromise with unbelief is so important. Linguistics, because all Study uses language, and many a fallacious argument or system achieves plausibility partly through erroneous views of how language functions (see 3.133 and 9.2). Logic, because it also is involved in all Study, and because people have too often ignored how different may be the standards of cogency for a covenant-keeper and a covenant-breaker (see Appendix 4). Probability theory, because of the pervasive use of or reference to probability both in sciences and in historical argument.

Fortunately, some progress has already been made in three of these fields. In apologetics, Cornelius Van Til's works have developed a consistently Christian stance (see 8.4). In linguistics, I judge that Kenneth Pike's work is the beginning of a Genuine Speculative Linguistics (see 8.6). In logic, I point to Dirk H. Th. Vollenhoven and Nicolass van der Merwe,[1] though I fear that their work may have to be revised because it is still entangled in some of the problematics of cosmonomism (see 8.2, 8.3, and chapter 11). In probability theory, I am aware of no adequate work.

7.2 The description of this book

It has been objected to Hegel that his system accounted for everything except Hegel. Similarly, one may object to Dooyeweerd's *New Critique* that it accounts for everything except how the *New Critique* can speak theoretically about God, law, and cosmos (see 8.211). I do not think that I have the same problem as these men. I can talk about what I have done. This book is, roughly speaking, Refined Boundary Evangelical Study, that is, Evangelical Philosophy or Philosophical Theology, with special interest in Philosophy of Science and Philosophy of Study.

At times, of course, I may have been Speculative, and thus not completely Evangelical. However, the person who really understands what I am saying, understands that I am saying little more than the Bible itself says. At the cost of more labor, I could have said it without my technical terms. The Bible says it better and more effectively, too. Does that mean that this book is superfluous? No. Because of sin, people still have trouble understanding the Bible and seeing the implications of the Bible for science. This book is intended to jar them into a better understanding and to provide some tools for seeing how to begin reforming science.

NOTES TO CHAPTER 7

1. Vollenhoven, *De noodzakelijkheid eener christelijke logica* (Amsterdam: H. J. Paris, 1932); *idem*, "Hoofdlijnen der logica," *Philosophia Reformata* 13 (1948), pp. 59-118; van der Merwe, *Op weg na 'n christelike logika* (Potchefstroom: Ph.D. thesis, University of Potchefstroom, 1958).

8. Appendix 1

EVALUATION OF PREVIOUS REFORMED PHILOSOPHY

This book has been devoted primarily to a *positive* exposition of Evangelical Philosophy. Hence, for the most part, I have not dealt with other positions, even with those claiming to be Christian. It seems only right, at this point, for me to give some account of why I have not depended more extensively on other Christian philosophies. This question becomes particularly pressing in the case of cosmonomic philosophy, since the questions that I am endeavoring to answer are closer to their questions than to those of apologists like Cornelius Van Til and (to a certain extent) Gordon Clark.

My answer in part is that I wanted to say different things, and in a different way, from what others have said. That should be obvious from the previous chapters. In particular, I wanted to demonstrate that valuable materials for Philosophy can be found in the Bible itself. But besides this, I judged that those philosophers closest to discussing the same questions I discuss have given unsatisfactory answers. So I include in this appendix a brief criticism (for the most part ontological criticism) of some selected figures.

8.1 *Non-Reformed philosophy*

First, let me say a word about the problems I find in general evangelical non-Reformed philosophy of science.

8.11 *Theological differences*

Theological differences prevent me from leaning too heavily on evangelical philosophy. In particular, lack of clarity about the sovereignty of God and the role of God's decrees forms a serious obstacle to good philosophy. Moreover, compromising syncretism in philosophy tends to follow in the steps of confusion about man's depravity.

8.12 Need for systematic treatment

Non-Reformed evangelicals have just not exhibited the same depth of concern for a systematic treatment of philosophy of science from a consistently Christian point of view.

8.13 Familiarity

I personally have not extensively studied evangelical non-Reformed philosophy.

Second, let us look at Reformed philosophy. I confine myself to the most outstanding contemporary figures: Herman Dooyeweerd, Hendrik Stoker, Cornelius Van Til, Gordon Clark, and Kenneth Pike. Obviously, a critique of any one of these men could form a book by itself; so I will confine myself to rather general remarks, without extensively substantiating them.

8.2 Herman Dooyeweerd[1]

I admire Dooyeweerd's genius and erudition. But I am also dissatisfied. I divide the expressions of dissatisfaction and objections into two categories: major and minor. Minor objections by themselves would be "nonfatal." Nevertheless, many of the objections are vaguely interrelated. I include also objections that may rest (as is often claimed by proponents of cosmonomic philosophy) on a misunderstanding of Dooyeweerd, but which at least show that Dooyeweerd has left himself open for all kinds of misunderstanding.

Minor Objections

8.201 Two "theoretical thoughts"

New Critique contains two disparate accounts of the nature of theoretical thought. These two accounts are not reconcilable with one another, and one of them (that theoretical thought is the setting in opposition of the analytical aspect to some nonanalytical aspect) is inconsistent with the existence of the disciplines of logic and philosophy.[2]

8.202 *The impossibility of neutrality*

New Critique claims to conduct an argument that can be followed independently of one's religious presuppositions, at the same time that its own conclusion denies the possibility of such a religiously neutral maneuver.[3]

8.203 *The supratemporal heart*

Scriptural teaching shows that the heart of man is not supra-temporal—contrary to Dooyeweerd's view.[4]

8.204 *The idea of law vs. the idea of creation*

The idea of law is not broad enough to form a completely adequate base for Christian philosophy. It must be supplemented with the idea of creation.[5]

8.205 *A chain of being*

Dooyeweerd's theme of the priority of unity to diversity (in law and cosmos) results in a kind of chain-of-being picture not unlike scholasticism.[6]

8.206 *Uninterpreted metaphor*

The picture of the *New Critique* rests on a series of uninterpreted metaphors that can prove just as misleading as helpful.[7]

8.207 *The naïve/theoretical distinction*

The naïve/theoretical distinction is difficult to maintain. This objection can take two forms. (a) There are those who maintain that thought is only relatively naïve or relatively theoretical, with a continuum in between.[8] (b) There are those who, so far as they know, have never "thought theoretically" in the way that Dooyeweerd describes.[9] Perhaps these two boil down to the same difficulty in the end, because it is difficult to interpret what Dooyeweerd says about the distinctiveness of theoretical thought in contrast to naïve experience.

8.208 *Assertion without demonstration*

Dooyeweerd never really justifies his way of looking at the world (compare 8.26). In particular, he leaves some people wondering about questions like the following. What is a modal aspect? Why are there listed just the ones that Dooyeweerd lists? What does it mean that one aspect is above another? How can we tell that one is above another?[10]

8.209 *Questionable practical results*

Application of the cosmonomic philosophy to practical problems has led to questionable results.[11]

8.210 *Problems with Genesis 1–2*

An unswerving adherence to the view that the human heart (or Adam?) is the "root" of the cosmos appears to require denying *a priori* that the apparent temporal order in Genesis 1 is real. Moreover, if all corruption in the cosmos proceeds from Adam's sin, it leaves no place for the temptation by the serpent *before* Adam's sin.[12]

8.211 *The mention of God "theoretically"*

New Critique speaks about God in a way that is *prima facie* inconsistent with its own strictures about the limitations of theoretical thought. This maneuver is too reminiscent of Kant to avoid suspicion.[13]

8.212 *Ignoring the Bible*

Because of its naïve/theoretical distinction, *New Critique* has not explored how far the Bible itself can give us a "philosophy."

8.213 *The problem of angels*

Are angels in the cosmos or not? The cosmos is sometimes described as what is subject to the law, but sometimes as what we have access to.

Major Objections

Under this heading I classify objections that either (a) are of great intrinsic importance to the Christian faith, or (b) if valid, can only be met by a radical reworking of the system, or (c) both. This is obviously a matter of *relative* major as opposed to *relatively* minor objections.

8.214 *Ethical Standards*

Does Scripture tell us all that we need to know about what human conduct pleases God (4.1), or are there extra-scriptural norms in addition? Dooyeweerd is unclear on this crucial question, and some of his followers take the wrong view.[14]

8.215 *Power and meaning of the word of God*

Dooyeweerd separates the power of the word of God from its meaning.[15]

8.216 *God-talk*

Talk about God becomes problematic in Dooyeweerd's system.[16] A special case of this is the problem of talking about the three persons of the Trinity.[17]

8.217 *A nondivine mediator*

The "law" functions as a not-fully-divine mediator between God and men.[18]

8.218 *Ambiguity and the "as such"*

The task that Dooyeweerd has set for himself of giving a critique of theoretical thought "as such," a critique to which unbelievers might agree formally, inevitably results in a document that is *radically* ambiguous between Christianity and anti-Christianity.[19] Even an appeal to the theme of "creation, fall, and redemption in Jesus Christ" cannot rescue it from ambiguity, because theological liberals, Barthians, Tillichians, etc., can use these terms with their own meanings.[20] Dooyeweerd does not distinguish his sense of 'creation,' 'fall,' 'redemption,' and 'Jesus Christ' from non-Christian senses.

8.219 *The sovereignty of God*

I must object against the disconcerting tendency in cosmonomic philosophy to assert the sovereignty of God in a highly ambiguous way, a way that leaves room for a virtually Pelagian view of the freedom of man. It is said that man must conform to the "norms" in the psychical (then why is psychology prescriptive?) and lower spheres, but that in the spheres above the psychical he may conform or disobey.[21] Does this imply, then, that God does not decree beforehand what *in particular* is going to happen as far as it relates to (say) the linguistic sphere? Then apparently Cyrus may or may not issue his decree (linguistically qualified? or historical? or juridical?), and either choice he makes will be under the "norms" of the juridical, linguistic, historical, etc., spheres.

This also illumines another problem, namely, that the "norms" or "laws" of cosmonomic philosophy appear to be only general laws ($F = ma$; you must speak grammatically, etc.), but *not* particular (Cyrus will issue the decree). If so, this philosophy is deistic. The fact that cosmonomic philosophy does not speak with *crystal clarity* on an issue as fundamental and life-transforming as this demonstrates that it cannot make a serious claim to be "Reformed" philosophy.[22]

8.3 *Hendrik Stoker*[23]

Stoker is free from some of the criticisms directed against Dooyeweerd. That will account for the greater similarity that my own exposition has to his. However, I have not found a clear distance between Stoker and Dooyeweerd on the matters of 8.207, 8.208, 8.212, 8.214,[24] 8.217, and 8.219. In regard to 8.217, Stoker is, if anything, even more explicit than Dooyeweerd about the nondivinity of the law:

> The cosmos has two sides . . . , namely cosmic 'things' (matter, plant, animal, man), which are subject to the cosmic law-order, and the cosmic law-order that holds for the 'things.'

> God created this law-order together with the cosmos. . . . The cosmic law-order is creaturely, non-self-sufficient, and wholly and completely in God's hand.[25]

8.4 *Cornelius Van Til*[26]

I have no *fundamental* objection to the substance of Van Til's work. However, I have several minor problems with it.

8.41 *Emphasis*

Van Til has confined himself to the root problems of apologetics.[27] I wish to complement this with a more positive exposition.

8.42 *Terminology*

Van Til's use of terminology taken over from non-Christian philosophy has opened the way for much rash criticism and misunderstanding of his view. I wish to develop some terminology that can say what he is saying in a less metaphorical or paradoxical way.

8.43 *Law*

It is not clear what Van Til thinks about "law" (see 8.214).

8.44 *Exegesis*

Van Til has generally avoided exegesis of the Bible (see 8.212).[18]

8.5 *Gordon Clark*[29]

My problems with Gordon Clark's work are as follows.

8.51 *Rationalism*

The stock accusation against Gordon Clark is that he is a rationalist, that he overemphasizes "reason." I basically agree with the objection, but as it stands it is exceedingly vague.[30]

8.52 *The law of contradiction*

Gordon Clark's striving after precision has, paradoxically, produced some rather confusing blur at other points in his system. What does he mean by "consistency"? What would it mean to test a system for consistency? It means using the law of contradiction.[31] But what is the law of contradiction? An attempt to spell this out will result either in a (question begging) Christian view of contradiction (cf. Van Til) or a platitudinous truth: "A statement cannot be both true and false at the same time and in the same way." But this platitude is not enough to show that non-Christian systems are contradictory.

It is not enough, that is, unless some kind of nontrivial content is once more poured back into 'true,' 'false,' or 'same way.'[32] Hence I judge that Gordon Clark's apologetics is an unstable equilibrium between Van Til and evidentialists.

8.53 Knowledge

Clark's theory of knowledge also has an odd blur. Suppose, on the one hand, that Clark means to use the word 'know' as an ordinary English speaker. Then his denial[33] that people have everyday knowledge not deduced from Scripture is in fairly obvious conflict with Genesis 3:7; I Samuel 4:6; II Samuel 11:16; 14:1, etc.

Suppose, on the other hand, that he is using 'know' in a special ("strict"?) sense. What could that sense be? The most obvious meaning of his word 'know' (call it 'know$_2$') is this: a person is said to know$_2$ that x if and only if God has said in Scripture x to that person, and the person has believed it. In that case Clark's statements about knowing$_2$ are trivially true, but unhelpful and confusing to his readers (who cannot forget the English 'know').

8.54 Language

Clark's view of language is rather simplistic.[34] His ideal seems to be that each word would have one precise meaning. But this ideal certainly does not come from Scripture in any obvious way. Where *does* it come from? Clark's own trenchant criticisms of the unwarranted assumptions of unbelievers now recoil upon himself.

8.6 Kenneth Pike[35]

Kenneth Pike has been completely ignored by other Christian philosophers, even though, in my judgment, he is the greatest living Christian Speculative Philosopher. The present ignorance, however, is to a large extent understandable. (a) Pike's lifetime work has been chiefly devoted to linguistics rather than to philosophy. He gives little more than pointers as to how his philosophical methods may be extended beyond linguistics. (b) He has not addressed himself directly to historical problems of philosophy, nor indeed does he claim much acquaintance with such problems.[36]

(c) Few of his writings have focused on Theological Philosophy, and the few that have, have dealt with only special issues. (Thus I call Pike a *"Speculative* Philosopher." Van Til is the far greater figure in Theological Philosophy.) (d) Pike's "Philosophy" is so different in both style and content from what is conventionally known as philosophy that it has not been recognized as answering philosophical questions.

NOTES TO APPENDIX 1

1. For my purposes, Dooyeweerd's most important works are *A New Critique of Theoretical Thought* (Philadelphia: Presbyterian and Reformed, 1969) and *In the Twilight of Western Thought* (Nutley, N. J.: The Craig Press, 1968). For extended bibliographies of Dooyeweerd, see Peter Steen, *The Idea of Religious Transcendence in the Philosophy of Herman Dooyeweerd* (unpublished; Chestnut Hill, Pa.: Westminster Theological Seminary, Ph.D. thesis, 1970), pp. v-xviii; and *Philosophy and Christianity; Philosophical Essays Dedicated to Professor Dr. Herman Dooyeweerd* (Kampen: Kok, 1965), pp. 449-452.

2. Hendrik van Riessen, *Wijsbegeerte* (Kampen: Kok, 1970), pp. 116ff.

3. *Ibid.*, pp. 124ff.

4. J. M. Spier, *Tijd en eeuwigheid* . . . (Kampen: Kok, 1953); Peter Steen, *The Supra-temporal Selfhood in the Philosophy of Herman Dooyeweerd* (unpublished; Chestnut Hill, Pa.: Westminster Theological Seminary, Th.M. thesis, 1961); Steen, *Idea;* John M. Frame, *The Amsterdam Philosophy: A Preliminary Critique* (Phillipsburg, N. J.: Harmony Press, c. 1972), pp. 22-25.

5. Hendrik Stoker, *Die wysbegeerte van die skeppingsidee* . . . (Pretoria: de Bussy, 1933). Cf. Dooyeweerd's response in *New Critique*, I, pp. 94-97. If this dispute involves only an Emphasizing Reductionism, it is not a significant objection.

6. Steen, *Idea.*

7. Frame, *Amsterdam Philosophy*, pp. 12ff. Such uninterpreted metaphors make Slippery Reductionism exceedingly easy in the cosmonomic system. See 9.2.

8. *Ibid.*, pp. 6-14. It looks as if van Riessen may be maintaining a continuum position in *Wijsbegeerte*, pp. 77ff.

9. Cf. Appendix 2.

10. Frame, *Amsterdam Philosophy*, pp. 20-22. Notice Hendrik Hart's remark, "But here again, the correctness of the point [the point of Dooyeweerd that to be conscious man must transcend time] is posited, not demonstrated" ("Problems of Time: an Essay," *Philosophia Reformata* 38 (1973), p. 37. See also Appendix 3.

11. Frame, *Amsterdam Philosophy*, pp. 40-49.

12. Noel Weeks, *Creation Themes in the Psalms* (unpublished; Chestnut

Hill, Pa.: Westminster Theological Seminary, Th.M. thesis, 1968), pp. 86-95.
13. Immanuel Kant maintains that the categories apply only to the "phenomenal world." Yet his own works use words of ordinary language to talk about the "noumenal world."
14. Frame, *Amsterdam Philosophy*, pp. 32ff.
15. Norman Shepherd, "The Doctrine of Scripture in the Dooyeweerdian Philosophy of the Cosmonomic Idea," *Christian Reformed Outlook* 21 (February, 1971), pp. 18-21; (March, 1971), pp. 20-23; Frame, *Amsterdam Philosophy*, pp. 32-40. Note also the discussion of related problems of hermeneutics in *ibid.*
16. *Ibid.*, pp. 25-27. Cf. 8.211.
17. Dooyeweerd's "numerical aspect" is an aspect of the "cosmos," but not of the "Origin." How, then, can we use the word 'three' in speaking of the "Origin"?
18. Frame, *Amsterdam Philosophy*, pp. 27-31; Weeks, *Creation Themes*, pp. 84-86. Cf. 8.205.
19. Cornelius Van Til, *Jerusalem and Athens*, ed. E. R. Geehan (Philadelphia: Presbyterian and Reformed, 1971), pp. 89-127.
20. Cf. Frame's remark, "Perhaps, indeed, the terms "creation," "fall" and "redemption," used in this sort of context (and sharply distinguished, as we have seen, from the *doctrines* of creation, fall and redemption) are mere codewords to designate that unnameable brute power. Perhaps it might even be possible to substitute "x," "y," and "z" for "creation," "fall" and "redemption" " (*Amsterdam Philosophy*, p. 34).
21. J. M. Spier, *An Introduction to Christian Philosophy* (Philadelphia: Presbyterian and Reformed, 1954), pp. 76f.
22. It may, of course, be claimed that the lack of clarity is an oversight. Well, it is a remarkably gaping oversight for a philosophy that claims that it is covering the field. Or it may be claimed that it is not the business of philosophy to make theological statements. This is probably the answer most consistent with cosmonomism, but it is convincing only for those who have accepted the cosmonomic radical distinction between philosophy and theology. What *are* those statements about law claiming to be? Has the philosopher no concern for the fact that regardless of what he *says* about the limits of his field, people being sinners will read the statements about law in a deistic sense unless he guards against this?
Or, as a third alternative, it may be claimed that the Reformed churches have in the past had a mistaken view when they said that God's law extended even to the point of saying that Cyrus would issue the decree. In that case, the issue needs open discussion rather than concealment.
23. For my purposes, Stoker's most important work is *Beginsels en metodes in die wetenskap* (Potchefstroom: Pro Rege-Pers, 1961). For an extended bibliography of Stoker, see Stoker, *Oorsprong en Rigting*, II (Kaapstad: Tafelberg, 1970), pp. 435-442.
24. Stoker's "self-evident" principles might be an additional source of norms for human conduct. See *Beginsels*, pp. 44-45.
25. "Die kosmos het twee kante . . . , nl. die kosmiese ,ietse' (stof, plant,

dier, en mens), wat aan die kosmiese wetsorde onderworpe is, én die kosmiese wetsorde wat vir die ,ietse' geld." "God het dit [wetsorde] saam met die kosmos gaskape. . . . Die kosmiese wetsorde is creatuurlik, onselfstandig en geheel en al in Gods hande" (Stoker, *Beginsels*, pp. 199, 203).

26. The works of Van Til most relevant to my purposes are *The Defense of the Faith*, 3rd ed. (Philadelphia: Presbyterian and Reformed, 1967); *A Christian Theory of Knowledge* (Philadelphia: Presbyterian and Reformed, 1969); *A Survey of Christian Epistemology*, In Defense of the Faith, vol. III (Philadelphia: den Dulk Christian Foundation, 1970); *The Dilemma of Education* (National Union of Christian Schools, 1954); *Christian-theistic Evidences* (Philadelphia: Presbyterian and Reformed, 1961); and *An Introduction to Systematic Theology* (Philadelphia: Presbyterian and Reformed, 1966). For an extended bibliography of Van Til, see E. R. Geehan, ed., *Jerusalem and Athens* (Philadelphia: Presbyterian and Reformed, 1971), pp. 492-498.

27. As Van Til himself admits in *Jerusalem and Athens*, p. 73.

28. But see Van Til, *An Introduction to Systematic Theology*, pp. 142-144.

29. The works of Clark most relevant to my purposes are *A Christian Philosophy of Education* (Grand Rapids: Eerdmans, 1946); *A Christian View of Men and Things* (Grand Rapids: Eerdmans, 1952); and *The Philosophy of Science and Belief in God* (Nutley, N. J.: Craig, 1964). For an extended bibliography of Gordon Clark, see Ronald H. Nash, ed., *The Philosophy of Gordon H. Clark* (Philadelphia: Presbyterian and Reformed, 1968), pp. 513-516.

30. It might simply be a case of Emphasizing Reductionism.

31. Clark, *Christian View*, p. 31.

32. I fear, however, that Clark and others who put much stock in the law of contradiction will have difficulty seeing this. For those who are skeptical about these claims respecting the law of contradiction, I include some further analysis of Clark in Appendix 4.

33. Clark, "The Axiom of Revelation," *The Philosophy of Gordon H. Clark*, ed. Ronald H. Nash, pp. 89ff. See also the further discussion by Nash (*ibid.*, pp, 173ff.) and George I. Mavrodes (pp. 227ff.).

34. See the exposition by Ronald Nash, "Gordon Clark's Theory of Knowledge," in *ibid.*, p. 129, and David H. Freeman, "Clark's Philosophy of Language," in *ibid.*, pp. 257-275, based largely on Gordon Clark, *Religion, Reason and Revelation* (Philadelphia: Presbyterian and Reformed, 1961). See also Clark, *Christian View*, p. 292. It may be that this is only an impression of mine, and that Clark's view of language is really less rigid (allowing, for example, for words with a certain range of meaning and "fuzzy borders").

35. The work of Pike most relevant to my purposes is *Language in Relation to a Unified Theory of the Structure of Human Behavior*, 2nd revised ed. (The Hague-Paris: Mouton, 1967). For an extended bibliography of Pike, see Pike, *Selected Writings to Commemorate the 60th Birthday of Kenneth Lee Pike*, ed. Ruth M. Brend (The Hague-Paris: Mouton, 1972), pp. 326-331.

36. However, this may actually be an advantage in terms of avoiding non-Christian formulations of the problems.

9. Appendix 2

THE NAIVE/THEORETICAL DISTINCTION

Now I wish to say something directly about Herman Dooyeweerd's naïve/theoretical distinction and the problems associated with it.[1] I intend to concentrate on Dooyeweerd because he has written the most extended and detailed account of the distinction. Nevertheless, I hope to pose certain questions and problems that bear on everyone who holds to such a distinction.

I will criticize Dooyeweerd's work by means of the method outlined in 3.133. It seems to me that Dooyeweerd's work exhibits Emphasizing, Exclusive, and Slippery Reductionism. Namely, he combines reduction of the Wave View to Particle and Field Views and reduction of Sapiential Study to Refined and Sensitive Study. Both of these reductions are accomplished by way of Dooyeweerd's naïve/theoretical distinction. If this be true, the method of 3.133 for dealing with Reductionisms should apply.

9.1 Ontological criticism

I have already engaged in "ontological" criticism (3.1331) of Dooyeweerd in 8.2. Way 1 of criticism, appealing directly to the language of the Bible, is used particularly in 8.203, 8.209, 8.210, 8.213, 8.214, 8.215, 8.216, 8.217, and 8.219. Way 2 of criticism, appealing to the Functions of God, is in 8.216.

Way 3 of criticism is to "agree" with Dooyeweerd but with a twist. Since I believe the naïve/theoretical distinction to be crucial, let us "accept" this distinction and then try to show that it does not say much. One way is to so broaden the scope of "naïve experience" that it includes most science.

For example, let us argue that this book and most theology books are works of naïve experience. This follows from some things that Dooyeweerd says about theoretical thought. Dooyeweerd says that "in the theoretical attitude of thought we analyze empirical reality by separating it into its modal aspects" (I, p. 38). And "in this process of theoretical thought, characterized by its antithetical attitude, every correct formation of concepts and judgements rests upon a sharp distinction among the different aspects of meaning and upon a synthesis of the logical aspect with the non-logical aspects of our experience which are made into a "Gegenstand" " (I, p. 18). Now, before Dooyeweerd, no one ever analyzed empirical reality by separating it into modal aspcts. Least of all did they do this "sharply." After all, no one before Dooyeweerd has had precisely *this* list of "aspects" with precisely *this* meaning. This book, for example, has a list of "Functions" and "Modes." But the "Functions" are not the same as Dooyeweerd's aspects, and hence I have not really analyzed things into aspects, and neither has any theologian (at least previous to the *New Critique*).

But more conclusive than this is something else that Dooyeweerd says about theoretical thought. He says, "We analyze empirical reality" (I, p. 38). Elsewhere it becomes clear that Dooyeweerd is speaking of *created* reality (I, p. 4). The "aspects" of which he speaks are "the fundamental universal modalities of *temporal* being" (I, p. 3, n. 1; italics mine), "the modal aspects of our cosmos" (I, p. 3). Theoretical thought is an analysis of Creation (or perhaps only the Cosmos). Dooyeweerd evidently wants to exclude the possibility that theoretical thought could deal with other things *besides* Creation. Therefore, insofar as this book or theology books discuss God, they are not theoretical thought. What Dooyeweerd says does not apply to them.

To a lesser degree the same is true of any science book. Because, as we have seen, science inevitably involves knowledge of God, and scientific statements say something about the Word of God, what Dooyeweerd says apparently does not apply to them either.

Of course, Dooyeweerd may claim that this book or other theology books are *really* "theoretical," even though they do not recognize it. In that case, let him *show* that they are "theoretical" in his sense.

If the quotes above are to be taken seriously, he can do so only by showing that this book and others really do not talk about God, but about the pistical aspect. But this is Exclusive Reductionism. Theology does talk about and study God—at least that is what we say in ordinary language. Perhaps Dooyeweerd wants to change that language, and say, for example, that theology talks about what people *believe* about God. Very well, but then it is talking about what people believe about *God,* and hence it is talking about God. Dooyeweerd might say, "Theology talks about what people believe with ultimate certainty." I reply, "Yes indeed, but it also talks about God." The only way in which Dooyeweerd might evade this is by trying to alter the meaning of 'theology,' or 'talk about,' or 'God,' but that will simply involve him in Slippery Reductionism.

Suppose now that "theoretical thought" is defined not in terms of a limitation of the subject-matter to the cosmos, but by the "Gegenstand relation." I answer, "Perhaps there are people who are constitutionally unable to have a Gegenstand relation to what they are studying—just as there are people constitutionally unable to roll their tongue into a semicylindrical shape or to wiggle their ears. Such people must get along as best they can without the advantages of this skill. I am inclined to think that I am one of these people (because I still do not know what a "Gegenstand relation" is)."

What, now, can Dooyeweerd say? He can, of course, assert that I am, definitely, thinking theoretically, but that I just have not noticed it. In that case, I am somewhat at a loss as to *how* he knows, unless he is an extraordinary telepath. That line of argument does not seem to be particularly fruitful. I would suggest that a more fruitful course would be to look at what I *say* (or write as the case may be). Surely Dooyeweerd is not really claiming that he knows what "thinking mechanisms" (rolling the tongue and the like) I use. What he is claiming that he reads what I write, and that what I write is "theoretical." Therefore, let him give us some criteria to distinguish when a piece of writing is "theoretical" and when it is naïve."

I already have a suggested criterion for him. A piece of writing is "theoretical" if it uses technical terms. The trouble is, terms can be *more or less* technical, and technical terms can occur *more or less* frequently in *more or less* crucial roles. (When the wife of a physicist says, "My husband is running the bevatron today," is she speaking very theoretically?) I am afraid that Dooyeweerd will find it exceedingly difficult to draw a hard-and-fast line between naïve and theoretical writing. In fact, on the basis of Kenneth Pike's study of language, it must be judged impossible rigidly to separate any part of language from everyday language or from "naïve experience."[2]

Is the naïve/theoretical distinction "sharp"? If, of course, Dooyeweerd himself is saying that in actual life there is a continuum, and that his "theoretical thought" never actually occurs "pure," the above objection falls to the ground. When I read the *New Critique* in my most sympathetic mood, I can sometimes think that this is what Dooyeweerd is saying. Is his term 'theoretical thought' itself, perhaps, a "theoretical abstraction" from the complex intertwinings of experience? Is "theoretical thought" a kind of ideal, pared-down picture of a situation that we may come into to various degrees?

I am attracted to this interpretation of Dooyeweerd by several considerations. First, (a) this interpretation would enable me to understand a good deal more of what he is saying. Also, (b) remarks here and there in the *New Critique* seem to point in this direction (particularly I, pp. 34, 40; III, p. 54). (c) Such an interpretation seems to me more consistent with Dooyeweerd's general dictum that "theoretical thought" abstracts from the fullness of naïve experience. Then presumably the theoretical term 'theoretical' in *New Critique* also has in it this quality of abstraction, elimination, and idealization.[3]

If that is the case, it quite alters my attitude toward and evaluation of the *New Critique*. Surely everyone is free to paint his own idealizations—which may be more or less useful for certain purposes. But then all the "musts" and "oughts" of Dooyeweerd—of which there are an astonishing number—are not necessarily justified.[4] No one need see any obligation to agree that Dooyeweerd's idealized picture of what science and philosophy and theology "ought to" be like is *the* ideal after which we should strive.

9.2 *Methodological criticism*

It is now time for us to look at the way in which Reductionism is involved in Dooyeweerd's *New Critique*. In the first place, Dooyeweerd himself admits to a reductionism of a sort in the Foreword of *New Critique:*

> On the basis of this central Christian point of view [that the "heart" is the religious root of human existence] I saw the need of a revolution in philosophical thought of a very radical character. Confronted with the religious root of the creation, nothing less is in question than a relating of the whole temporal cosmos, in both its so-called 'natural' and 'spiritual' aspects, to this point of reference [I, p. v].

So Dooyeweerd promises us a "Reduction" to the heart of man. Now there is no reason why this might not take place in the form of Emphasizing Reductionism, and so be harmless. But the *fact* that Dooyeweerd claims to accomplish such a Reduction furnishes us with no guarantee that Exclusive Reductionism or Slippery Reductionism is not involved. It is no easier to avoid sin in terms of a Reduction to the "heart" than it is in terms of any other kind of Reduction. Men may know that they have a "heart," and know that "out of the heart are the issues of life," and yet still not acknowledge that their own heart is sinful.[5]

From other places we can gather what role Dooyeweerd wants this "heart" to play in his system. He says, "In this whole system of modal functions of meaning, it is I who remain the central point of reference and the deeper unity above all modal diversity of the different aspects of my temporal existence" (I, p. 5). Thus 'I' is evidently equivalent to 'heart' as the central point of reference. Dooyeweerd continues,

> To be sure, the ego is actually active in its philosophical thought, but it necessarily transcends the philosophical concept. For, as shall appear, the self is the *concentration-point of all* my cosmic functions. It is a subjective *totality* which can neither be resolved into philosophical thought, nor into some other function, nor into a coherence of functions. Rather it *lies at the basis* of all the latter as their presupposition [I, p. 5].

'Ego' and 'self' appear to be other terms equivalent to 'heart.' Again,

> it [philosophic thought] issues from our own selfhood, from the root of our existence. This restlessness is transmitted from the

selfhood to all temporal functions in which this *ego* is actually operative. Inquietum est *cor* nostrum et mundus in corde nostro! [I, p. 11].

Hence Dooyeweerd wants to relate philosophy to self-reflection: "Γνῶθι σεαυτόν, "know thyself," must indeed be written above the portals of philosophy. But in this very demand for critical *self-*reflection lies the great problem" (I, p. 5).

But studying knowledge in relation to oneself is just a special case of studying knowledge in relation to men. Hence it falls under my "Refined Study." Thus in this way Dooyeweerd emphasizes Refined Study of a certain type. On the other hand, Sapiential Study is minimized, since Dooyeweerd is reluctant to admit that we have "theoretical" knowledge of God. Dooyeweerd is working with an Emphasizing Reductionism.

Dooyeweerd executes a further Reduction when he emphasizes thinking methods that are "sharp":

> In this process of theoretical thought, characterized by its antithetical attitude, every correct formation of concepts and judgements rests upon a sharp distinction among the different aspects of meaning and upon a synthesis of the logical aspect with the non-logical aspects of our experience . . . [I, p. 18].

Hence the Wave View (the View that focuses on fuzzy boundaries: 3.123) is minimized.

But having identified this Reductionism in Dooyeweerd, it is now fairly easy to go through the *New Critique* page by page and show that it is constantly engaged in *Slippery* Reductionism, utilizing the above Emphasizing Reductionism. *New Critique* is in fact a veritable masterpiece of Slippery Reductionism. This Reductionism takes place by means of an oscillation on Dooyeweerd's part between a broader, "naïve" meaning of his terms and a narrower, "theoretical" meaning—hence the absolute centrality of the naïve/theoretical distinction to his whole system. The "theoretical" meaning involves an elimination of, or rather minimizing of, Sapiential Study (emphasizing Refined Study and the self), and a minimizing of the Wave View.

I have not the space here to go through the *New Critique* from one end to the other, but I can perhaps hope to make my point by means of a few examples. The oscillatory ambiguity between naïve and theoretical meaning is already set up on the first page of the *New Critique,* which introduces terms like 'naïve,' 'theoretical,' 'analysis,' 'numerical,' 'spatial,' 'mathematical movement,' 'physical,' 'organic life,' 'feeling,' 'logical,' 'historical,' etc. All these terms have meaning in English (call it meaning$_1$, a nontechnical or "naïve" meaning) and a special meaning as technical terms in Dooyeweerd's philosophical vocabulary (call it meaning$_2$ or "theoretical" meaning).

Dooyeweerd admits that such a distinction between meanings is involved when he says that in naïve experience "we do not become aware of the modal aspects unless *implicitly.* The aspects are not set asunder . . ." (I, p. 38). Hence the above terms as used in "naïve experience" have a vagueness to them.[6] However, in "theoretical thought" Dooyeweerd wants to use the same terms "sharply (no Wave View) and reflectively (in a form of Refined Study). Hence he has an ambiguity in every technical term that he uses.

Of course, a certain amount of ambiguity in a system may be tolerable. One can be ambiguous without being in error. But in the *New Critique* the ambiguity is exploited in order to draw invalid conclusions. For example, consider the sentence "now philosophy should furnish us with a theoretical insight into the inter-modal co-herence of all the aspects of the temporal world" (I, p. 4). First, read that sentence with every term having meaning$_1$. It is a debatable statement. Some people might say that philosophy should lead us to mystical union with the absolute. I myself might say that philosophy should serve the kingdom of God, whether or not it "furnishes us with a theoretical insight. . . ." However, I would not be terribly unhappy about it furnishing us with such insight, because no doubt it would be a useful kind of insight to have. Likewise, many secular philosophers would be willing to agree to the statement.

Second, read the same sentence with meaning$_2$. Then it is *definitely* true, because Dooyeweerd virtually defines philosophy$_2$ as having this job.

Third, read the sentence giving 'philosophy' the meaning philosophy$_1$ and giving all other terms meaning$_2$. Then it is a demand that all other philosophers conform to the way that Dooyeweerd wants to work. It says that philosophy$_1$ should become philosophy$_2$. This statement is now an Exclusive Reductionism, accomplished by means of ambiguity.

As the next example, take Dooyeweerd's statement a little further on that "philosophical thought in its proper character, never to be disregarded with impunity, is theoretical thought directed to the *totality of meaning* of our temporal cosmos" (I, p. 4). Again, one could insert 'philosophy$_1$' or my term 'Philosophy' which is close to 'philosophy$_1$' in meaning. One could also insert theoretical$_1$ or my term 'Refined' which is close to 'theoretical$_1$.' In neither case would I be comfortable with the resulting sentence, because it appears to imply a *restriction* to "totality$_1$" and "temporal$_1$" ("created$_1$"?) reality. Can we not study God and his works? Moreover, some philosophers$_1$ do not accept that restriction. The sentence becomes still less acceptable if we read "totality$_2$ of meaning$_2$," because then it virtually requires conformity to Dooyeweerd's system. If, on the other hand, we read the sentence completely with meaning$_2$, it is again true by Dooyeweerd's definition of terms.

Another example:

> In this process of theoretical thought, characterized by its anti-thetical attitude, every correct formation of concepts and judgements rests upon a sharp distinction among the different aspects of meaning and upon a synthesis of the logical aspect with the non-logical aspects of our experience which are made into a "Gegenstand" [I, p. 18].

With the meaning theoretical$_2$, this statement is trivially true, because it describes the ground rules for doing theoretical$_2$ thought. But will everyone agree that this is a requirement for theoretical$_1$ thought? Or for Refined Study?

Next I take a crucial example from the second volume of *New Critique*. Dooyeweerd says,

> It will be clear why the ambiguity [!] in the pre-scientific use of terms does not concern us in this context. Our inquiry exclusively refers to the modal structures of meaning [II, p. 61].

It is ironic that Dooyeweerd should denominate pre-scientific use of terms as "ambiguous" (presumably, that is, ambiguous between various modal meanings). It is his *own* use that is really "ambiguous." But this is another case where there is one usage "ambiguous$_1$" in ordinary language and another usage "ambiguous$_2$" by Dooyeweerd. Let us say that terms are ambiguous$_2$ when they can be used in speaking of more than one aspect$_2$.

Pre-scientific use is in general ambiguous$_2$ but not ambiguous$_1$. Granted that homonyms exist, the Variation and vagueness in meaning of words is better called just that—Variation and vagueness—rather than ambiguity$_1$. Ambiguity$_1$ suggests a situation where, by verbal sleight of hand, deceitful or fallacious arguments can win easy acceptance.

Now doubtless "common parlance" is ambiguous$_2$. But by using the term 'ambiguous' ambiguously$_1$, Dooyeweerd suggests also that common parlance is ambiguous$_1$. Moreover, he suggests that his own terminology—the only terminology that is non-ambiguous$_2$—is non-ambiguous$_1$. Then his own system would really be "better" or less prone to fallacious argument than are other philosophical systems or than is common parlance. The shift from ambiguous$_1$ to ambiguous$_2$ enables Dooyeweerd to beg the question of whether *his* terminology is really the kind that avoids fallacy.

As the next example, I take an argument dealing more specifically with one aspect$_2$ (or is it aspect$_1$?).

> Every attempt to reduce the modal meaning of the latter [natural numbers] to purely logical relations rests, as will appear, on a confusion between numerical analogies in the structure of the analytical relations and the original kernel of numerical meaning. The latter can be found in nothing but quantity (how much) disclosing itself in the series-principle of the numerical time-order with its $+$ and $-$ directions. This modal time-order itself is determined by the quantitative meaning of this aspect [II, p. 79].

If these sentences are read with meaning$_2$, all is well. As a matter of fact, the above statements are virtually tautologies, because the terminology of "analogy$_2$," "kernel$_2$," and "time-order$_2$" virtually presupposes what is to be shown. On the other hand, if meaning$_1$ is substituted at points, the above sentences beg the question. For instance, shall we or shall we not agree that "the latter [the original kernel$_1$ or numerical$_1$ meaning$_1$] can be found in nothing but quantity$_1$ [or quantity$_2$?] (how much) disclosing itself in the series-principle of the numerical$_2$ time-order$_2$ with its $+$ and $-$ directions"? According to (say) Bertrand Russell, the answer is no. This is simply because Russell's view is not the same as Dooyeweerd's. On the other hand, if meaning$_1$ is substituted all the way through the last two sentences quoted above, Russell might conceivably agree because neither of the sentences talks about anything but "logic$_{Russell}$" (that is, "logic" in Russell's sense). Hence Dooyeweerd "wins" the argument with Russell and others by imposing his meanings$_2$ as the "right" ones.

As a final example, take Dooyeweerd's use of the term 'theology.' When Dooyeweerd says that "theology" is the science that studies the pistical aspect (II, p. 562), he appears to use the terms 'theology' in a sense close to what the world calls "science of religion."[7] Of course, Dooyeweerd has a special meaning for 'religion,' so that he himself would not use the phrase 'science of religion.' At any rate, let us call Dooyeweerd's technical sense of "theology" theology$_2$." This is, according to Dooyeweerd, "theology in its scientific sense."[8] But in this case, as always, Dooyeweerd does not say when he is using 'theology' as a technical term and when he is using it in an ordinary way. Apparently he thinks that he can do both. Hence his argument is muddled.

For instance, Dooyeweerd says,

> But, as such, it [the central motive of the Holy Scripture] can never become the theoretical object of theology; no more than God and the human I can become such an object.[9]

If meaning$_2$ is used in this sentence, it is true by Dooyeweerd's definition. If meaning$_1$ is used for 'theoretical' and 'theology,' it begs the question. It might be supposed from Dooyeweerd's *own* strictures that only meaning$_2$ is involved:

> If we wish to succeed in positing the problem concerning a Christian philosophy and its relation to dogmatic theology in a clear way, we must in the first place avoid any ambiguity [!] in the use of the terms and define [!] what we understand by them.[10]

Unfortunately, it is clear later that Dooyeweerd blithely goes on to apply what he has said (supposedly about theology$_2$) to theology$_1$. Namely, he applies his conclusions to Luther, Calvin, Kuyper, and "Reformed theology."[11]

Of course, one could get out of the muddle by interpreting Dooyeweerd to mean that what people ordinarily call "theology" is bad, and that the people who are now doing theology$_1$ (ordinary theology) should be doing theology$_2$ (science of the pistical aspect). So far so good. But why is theology$_1$ bad? Dooyeweerd does not have a clear answer to this. If he were to say that theology$_1$ misrepresents biblical teaching, that would be a clear answer. The remedy would be to show from the Bible where it is deficient. What Dooyeweerd apparently claims, however, is a little different. Namely, he claims that theology$_1$ *must* misrepresent Scripture, because of its confused methodology.[12] Note the difference between these two claims.

The reason why theology$_1$ *must* turn out bad is that one must either study the pistical aspect theoretically *or* receive naïve religious knowledge of the Word of God. Since theology$_1$ is neither of these, it is confused. But we are confronted with still a further problem. Suppose that we look at the claim, "One must either study the pistical aspect theoretically *or* receive naïve religious knowledge of the Word of God." What does 'must' means? Does it mean that there is no possible third alternative (such as, e.g., trying to find out what the Bible teaches about adultery in order to conform one's life to its teachings)? Or does it mean that one "ought not" to pick a third alternative?

If we give the second of these two answers, what basis is there for the "ought"? Someone should attempt to convince us *from Scripture* that such an "ought" exists. A person who does not have the patience to try to convince us from Scripture is setting himself up as an arbiter of Ethics, and that is rebellion against the ethical sovereignty of God.

Suppose, on the other hand, that we give the first of the two answers. Then theology$_1$ does not exist (a rather paradoxical conclusion). Or, alternatively, theology$_1$ is simply a combination of theology$_2$ and "naïve religious knowledge," in which case it is a combination of two kinds of knowledge each of which is all right, and so (unless it is a "bad" combination—but what would that mean?) theology$_1$ is all right. But then Dooyeweerd does not succeed in criticizing anything.

But let us try once more. Suppose that theology$_1$ is a "bad" combination because when it makes the transition from naïve religious knowledge to theology$_2$ it does so unreflectively. But why should lack of reflection at this point be bad? Is it because theology$_1$ thereby opens itself up to some error of equivocation? But then let Dooyeweerd demonstrate and not merely claim (Exclusive Reductionism?) that whatever "equivocation" is involved must lead to error or distortion.

9.3 Axiological criticism

In this final section I shall look at the *New Critique* more closely from the standpoint of its purpose. After reading the previous section (9.2), a reader may be disposed to ask, "Can it really be that simple? Can Dooyeweerd really have made such obvious terminological mistakes? Surely Dooyeweerd must mean something else." I hope that he does mean something else, but I must go by what I read and not by what I might have hoped that he wrote.

But actually, that Dooyeweerd should have fallen into such errors is not surprising. Slippery Reductionism has been common in Western philosophy almost from the beginning. Dooyeweerd has followed in the tracks of his predecessors.

9.31 Dooyeweerd and Western philosophy

For instance, I have said that Dooyeweerd engaged in minimizing the Wave View and in emphasizing self-reflective Refined Study. Neither of these two Reductionisms is new. First, minimization of the Wave View has been a philosophic practice ever since Aristotle's logic and grammar. Ideally, implies Aristotle, terms should be "sharp." Of course, minimizing the Wave View may take place in scientific practice purely as Emphasizing Reductionism, and so be harmless. In fact, the possibility of human precision using the Particle View is based on the fact that God knows everything exhaustively, with perfect precision.

But then the temptation is present to claim that one's own human precision is godlike precision, that the Wave-minimized view is a more "ultimate" explanation, that it is an intrinsically "better" and "deeper" explanation of things. This is Exclusive Reductionism, and by the ambiguous use of terms it easily slides over into Slippery Reductionism. Such Reductionism is used to sustain the illusion of godlike power and authority (note the "musts" and "oughts" derived by philosophers from Slippery Reductionism). This power and authority supposedly derive from superior understanding.

Second, what about the emphasis on self-reflective, Refined Study? Self-reflection is as old as Socrates, and receives continued attention in Kant, existentialism, and phenomenology. Dooyeweerd claims to be new in introducing the "heart." But as we saw in 9.2, a perusal of the *New Critique* quickly shows that 'self,' 'I,' 'ego,' 'selfhood,' 'central selfhood' are used as virtual equivalents of 'heart.' Once again, Dooyeweerd is operating with an ambiguity between 'heart$_2$' as a technical term for selfhood$_2$ or ego$_2$ and the biblical ("naïve$_2$") word 'heart$_1$.'[13]

The possibility of relating all knowledge to self-knowledge is based on the fact that what one knows is that which *he* knows: everyone is involved in his own knowledge. Moreover, self-knowledge and knowledge of God are interrelated. But suppose, in the extreme case, that it is said that all knowledge is "really" self-knowledge. Then Reductionism to self-knowledge becomes Exclusive Reductionism; it is a godlike claim to self-sufficiency. Such a claim is an illegitimate imitation of God. In knowing himself, God knows what he will do, and hence knows everything.

The combination of the two Reductionisms—minimizing the Wave View for "precision," and emphasizing self-reflection—can be especially useful to the sinner. By means of minimizing the Wave View the sinner claims godlike knowledge over what he can and cannot be; by means of self-reflection he claims that such knowledge has its "source" (note the ambiguity) in himself. Hence by his own self-regulation of *what* he knows, he can always succeed in doing what is right in his own eyes. In other words, he plays the part of an "autonomous" man.

This happens in Kant, in Husserl,[14] and, despite his laudable intentions, in Dooyeweerd. Dooyeweerd himself has a name for these two Reductionisms.[15] Minimizing the Wave View is what Dooyeweerd calls the "science ideal." Emphasizing self-reflection is what Dooyeweerd calls the "personality ideal." The combination of the two he calls the "nature and freedom" motive. The sad, sad thing is that, in spite of all his Christian zeal, Dooyeweerd has simply invented a new form of what he criticizes.

9.32 *Dooyeweerd and my Philosophy*

I should now like to ask how Dooyeweerd and other philosophers using a "sharp" naïve/theoretical distinction would criticize my Philosophy. Dooyeweerd is hopeful that by means of his philosophy communication between philosophical schools may be enhanced.[16] His critique is designed to communicate with "immanence philosophy." For the sake of argument, let us suppose that it can do this. But can it communicate with me?

I think not. For suppose that Dooyeweerd tries to "communicate" with me "theoretically." That is, suppose that he tries to bring in his special terminology. Then I will proceed to take the terminology apart much as I have done in 9.2. The only way that Dooyeweerd can avoid this result is by removing the foundation on which I rest my criticism of his philosophical methodology. But I claim that this foundation is basically biblical and Exegetical.[17] So sooner or later Dooyeweerd will have to engage in exegesis with me, and try to show where I have misinterpreted and misapplied Scripture.

Now I welcome any such confrontation. I am not infallible. Perhaps Dooyeweerd can correct me. But then Dooyeweerd would be using the linguistic details of a "naïve" document in order to try to revise my philosophical foundations. Hence in that very moment he would, it seems, have to abandon the idea that "the Bible does not provide us with philosophical ideas."[18]

I hope that I am wrong on this point. I hope that, contrary to the apparent tenor of his statements, Dooyeweerd will in fact be able to engage in exegesis in order to help decide philosophical controversies.[19] But suppose not. Suppose that Dooyeweerd or someone else operating with a naïve/theoretical distinction refuses to lodge exegetical objections. Then such a person is acquainted with a body of teaching (namely, this book) purporting to be biblical, against which he raises only question-begging objections. Yet he refuses to admit that this book could *possibly* be the tenor of biblical teaching. I have a name for such refusal: hardening of the heart. It is this that John Frame fears when he writes,

> It may be wrong to say that the doctrine of Scriptural authority is the "most important" doctrine of our faith; but it is certainly true that if this doctrine is rejected, no other doctrine can be established. We believe that the approach of some Amsterdam philosophers to Scriptural authority which we have discussed above in fact eliminates that authority in the historic sense and elevates human reason as the ultimate rule for Christian faith and life. We reject that position in the strongest possible way.[20]

NOTES TO APPENDIX 2

1. The difference between naïve experience and theoretical thought is explained by Dooyeweerd as follows: "In the theoretical attitude of thought we analyze empirical reality by separating it into its modal aspects. In the pretheoretical attitude of naïve experience, on the contrary, empirical reality offers itself in the integral coherence of cosmic time. Here we grasp time and temporal reality in typical total-structure of individuality, and we do not become aware of the modal aspects unless *implicitly*. The aspects are not set asunder, but rather are conceived of as being together in a continuous uninterrupted coherence"—*A New Critique of Theoretical Thought* (Philadelphia: Presbyterian and Reformed, 1969), I, p. 38. See also the cross references listed under "naïve experience" and "theoretical thought" in *ibid*, IV, pp. 164-166, 238-239. For problems of interpreting this description, see John M. Frame, *The Amsterdam Philosophy: A Preliminary Critique* (Phillipsburg, N. J.: Harmony Press, c. 1972), pp. 6-14.

In the remainder of this appendix *New Critique* will be cited by volume and page number.

2. Kenneth Pike, *Language in Relation to a Unified Theory of the Structure of Human Behavior*, 2nd ed., revised (The Hague-Paris: Mouton, 1967), pp. 25ff. In particular, Pike says, "All psychological processes, all internal structured responses to sensations, all of thinking and feeling, must also be considered as parts of human behavior which will become structurally intelligible only when a theory, a set of terms, and an analytical procedure are provided which deal simultaneously and without sharp discontinuities with all human overt and covert activity. Language is but one structured phase of that activity" (*ibid.*, p. 32).

3. Cf. Dooyeweerd's remark, "This abstraction from the actual, entire ego

that thinks may be necessary for formulating the concept of philosophical thought. But even in this act of conceptual determination it is the self that is actually doing the work" (*New Critique*, I, p. 5).

4. "Now philosophy *should* furnish us with a theoretical insight into . . ." "Philosophy *should* make us aware, that . . ." "Philosophy *must* direct the theoretical view of totality over . . ." ("Philosophical thought in its *proper* character, never to be disregarded with *impunity*, is theoretical thought directed to . . ." (I, p. 4). "Can philosophy—which *ought to* be guided by . . ." (I, p. 5). "A philosophy which . . . *must* from the outset fail . . ." (I, p. 5). "In my central selfhood I *must* participate in the totality of meaning, . . ." (I, p. 8). "All *genuine* philosophical thought has therefore started as . . ." (I, p. 9). Italics are mine. Cf. 8.214.

5. In my opinion, this already indicates that Dooyeweerd's " 'Copernican' revolution" "of a very radical character" (I, p. v) is simply one more new way of looking at things, one more new perspective. Whether or not it is "Christian" must be determined by its more specific statements. Dooyeweerd is not "intrinsically" more Christian because of the particular focus on the heart that he adopts (to claim so would be Exclusive Reductionism).

6. Cf. Dooyeweerd's statements that "it will be clear why the ambiguity in the pre-scientific use of terms does not concern us in this context. Our inquiry exclusively refers to the modal structures of meaning. Pre-theoretical experience does not explicitly distinguish the modal aspects as such; it conceives them only implicitly within the typical total structures of individuality. Therefore pre-theoretical *terms* are not the subject of our present inquiry" (II, p. 61; italics mine).

7. Cf. also Dooyeweerd, *In the Twilight of Western Thought* (Philadelphia: Presbyterian and Reformed, 1960), p. 143.

8. *Ibid.*, p. 113.

9. *Ibid.*, p. 125.

10. *Ibid.*, p. 120.

11. Cf. *ibid.*, p. 145, and elsewhere. The psychological reason for this mistake is fairly simple. When a technical term is introduced (like 'mass,' 'force,' and 'action' in physics), it is presumably with the purpose of using the term from then on in the technical sense. For this reason a newly coined or contentless term like 'blip,' 'krad,' or even 'x' would serve just as well. The only reason for using an ordinary English term is that it provides a preliminary mneumonic device to help associate the term in some vague way with its definition. But the mneumonic device creates a danger: the uninitiate will still associate the term with its ordinary-language connotations.

However, in the case of Dooyeweerd's terms, this device and this danger are necessary to the system. If, for example, Dooyeweerd coined a term 'pisteology' or 'krad' instead of using the term 'theology,' the ordinary language connotations would not be available. Then ambiguity would not be so easy either psychologically or literarily. The loss of ambiguity would destroy the plausibility of the system.

12. Cf., for example, *ibid.*, p. 115: "The fatal step of confusing theoretical Christian theology with true knowledge of God."

13. Of course, Dooyeweerd is somewhat different from much secular philosophy in attributing to the heart a religious direction either for or against the Origin. But some secular philosophers might not really be unhappy with such language, especially if they were permitted to give their own sense to the word 'Origin.'

14. Note Dooyeweerd's statement: "Originally I was strongly under the influence first of the Neo-Kantian philosophy, later on of Husserl's phenomenology. The great turning point in my thought was . . ." (I p. v).

15. Of course, at this point as at all others my terms do not coincide with Dooyeweerd's technical terms, because Dooyeweerd's technical terms are weighted with the whole baggage of Dooyeweerd's system. Therefore readers will have to take my comparison between "minimizing the Wave View" and the "science ideal," and the other comparisons with a grain of salt. Nevertheless, it seems to me fair to point out the similarity between what Dooyeweerd senses is wrong with "immanence philosophy" and what is wrong with his own work.

16. "This critique has been presented as the only critical way of communication between a really reformatory Christian philosophy and philosophical schools holding in one sense or another to the supposed autonomy of theoretical thought"—Herman Dooyeweerd, "Cornelius Van Til and the Transcendental Critique of Theoretical Thought," *Jerusalem and Athens,* ed. E. R. Geehan (Philadelphia: Presbyterian and Reformed, 1971), p. 74.

17. Doubtless Dooyeweerd would object to the terms 'Exegetical.' It makes him think of theoretical$_2$ exegesis. But I mean studying and discussing what the Bible says (see 6.123). Of course, I do not get every single thing that I say in this book from exegesis. At the very least, I use lexicons, concordances, grammars, etc., which do not themselves derive wholly from Scripture. Moreover, in talking about the *New Critique,* I have had to know something about the *New Critique.* That knowledge also did not come from Scripture. But my discussion of knowledge (chaptre 5) implies that scriptural teaching itself warrants using all kinds of sources.

18. Dooyeweerd, "Van Til," p. 82. But the words 'idea' and 'philosophical' have a somewhat special sense in Dooyeweerd.

19. Dooyeweerd does occasionally quote Scripture. And he hints in *Twilight,* p. 148, that philosophy has *some* exegetical competence. However, it is not clear to me how far exegetical work is intended as a demonstration of some of his philosophical notions, and how far it is intended as a mere illustration or confirmation. A further deep problem arises in connection with Dooyeweerd's dictum that "when we speak of creation, we use human *words* varying with the language of which we avail ourselves, and multivocal in common parlance. But in biblical usage they have got an identical revelational meaning in so far as they relate to God in his self-revelation as the absolute Origin of all that through his Word has been called into being. This revelational meaning transcends every human concept . . ." ("Van Til," pp. 85f.). Here Dooyeweerd separates biblical language from all other language in a way that might call in question *all* exegesis.

20. Frame, *Amsterdam Philosophy,* pp. 39-40.

10. Appendix 3

ASPECTS

What is an "aspect" or "mode"? I have been troubled by the apparent arbitrariness of Dooyeweerd's list of fifteen aspects. Why these and only these? Why in this particular order?[1] Dooyeweerd makes no attempt to build up to the aspects by argument, but simply hands them over full-grown on the first page of *New Critique*. Granted that some kind of sense can be given to his selection, it is nevertheless true that *other* arrangements of the aspects have plausibility as well.[2] How can Dooyeweerd or others argue that their arrangement and selection is superior to others?

One way of experimenting with different arrangements is to treat aspects as "boxes" into which English adjectives (or modifiers in some other language) are classified. If aspects are a sort of division "innate" in the world, we should perhaps find adjectives falling into one category or another. This is because language is a gift from God equipping man to talk about God's Creation. If Creation had "aspects" in the cosmonomic sense, surely these aspects ought to be mirrored in some fashion in language. Since the "aspects" of cosmonomic philosophy are in some ways "adjectival" in nature,[3] a classification of adjectives might appear to be one method of exploration. (On the other hand, because of "sphere universality," one should expect many cases where adjectives are used in a quasi-metaphorical sense.)[4]

Of course, I am not suggesting that cosmonomic philosophy itself views aspects as boxes of adjectives. No. The leading cosmonomic thinkers would probably frown on the idea. My proposal is that, if the cosmonomic idea of "aspect" has become a mystery to a person, defining "aspects" in terms of groups of adjectives becomes one possible alternative to try. It becomes also an avenue by which some less metaphorical meaning might be assigned to the word 'aspect' or to the "order" of aspects.

This, at least, was my hope when I first looked at adjectives, and later verbs. However, my own work with adjectives and verbs has left the situation about as confused as it was before, as the reader can see. This appendix is essentially a report of a failure to find any pattern in language itself that could justify a particular "aspect" system in detail.

As a starting corpus of adjectives I have taken all adjectives from a randomly selected popular text reporting on a non-Western culture.[5] This particular type of text was chosen with the hope that it would include a significant sampling of adjectives from the "higher spheres," without disproportionate weighting toward one "sphere." I excluded from consideration verbal participles, nouns functioning as close-knit modifiers, possessive pronouns, 's-formations, hyphenated expressions, and proper (capitalized) adjectives. I recorded the adjectives on slips of paper and then shuffled them about into the most appropriate "aspect" piles.

The results were that, for *any* of the aspect schemes that I tried, (a) some adjectives fell more or less clearly into one pile, and (b) a troublesomely large number of adjectives were mildly ambiguous between two or even three aspects. In detail, the results are in Table 12 (using Dooyeweerd's modal aspects) and Table 13 (using my Modes and Functions). Table 13 needs some explanation. The Mathematical Function is subdivided into "Spatial," "Quantitative," and "Aggregative" parts. It is possible that such a subdivision is artificial. However, it is also possible that it may be related to the Particle, Wave, and Field Views of Mathematics, as follows.

Description. The *Aggregative, Quantitative,* and *Spatial Subfunctions* are, respectively, those parts of the Mathematical Function that the Particle, Wave, and Field Views focus on.

We can describe studies of these fields as follows.

Description. *Set Theory, Arithmetic,* and *Geometry* are, re-

Table 12

Dooyeweerd's modal scale classifies adjectives

---

pistical:
ethical: polygamous (or social?), bad
juridical: right, national, colonial, royal, independent (or social?)
aesthetic: magnificent, clean (or physical?), pleasant (or psychical)
economic: precious, rich, wealthier, pastoral (or historical?)
social: own (economic? juridical?), heathen, private, ethnic, customary,
    traditional (or historical?), famous, acceptable (or ethical?), unmet (?)
linguistic: representative (or historical? or analytical?)
historical: historical, new (or kinematic?), old (or kinematic?), difficult,
    ancient
analytical: simple, ultimate, profound (or linguistic?), special, paramount
psychical: humble (or ethical?), lazy, shy, curious, aloof (or social?), proud
biotic: elderly (or historical?), young, flowery
physical: corrugated (or spatial), fenceless, roast, hard, bitter, fragile, hot,
    inland (or spatial), stony, coastal, spotless, dark, orange, green, blue,
    white, black, brown, substantial (?)
kinematic: earlier (or historical?), steady
spatial: plump, small, thin, high, great, open, colossal, distant, upper, northern
    (or historical?), wide, huge, deep, highest
numerical: thousand, two, 10,000, 23 million, 100,000, 500, one, all, ten,
    uncounted, second, three, 112, fewer, dozen, 13,000, 3,900,000, every,
    four, many, other (or analytical?), double, more, much

(Each aspect is followed by adjectives apparently falling under it. One of the
problems is distinguishing historical and kinematic.)

---

spectively, the studies of the Aggregative, Quantitative, and Spatial Subfunctions.

See Table 14.

What about still other systems of aspects? Stoker's aspects are very similar to Dooyeweerd's.[6] But he eliminates Dooyeweerd's "historical" and "kinematic." The adjectives in these two aspects could go partly under physical ("earlier, steady, new, old, ancient"), partly under analytical ("difficult [?] historical [or social?]"). It seems more likely that Stoker would eliminate these adjectives from aspects altogether and put them under the "dimension of events."[7]

Seerveld's proposed modal scale differs from Dooyeweerd's mostly

Table 13

Modes and Functions classify adjectives

Personal:
  Sabbatical: bad, right, acceptable (?)
  Dogmatical:
  Presbyterial: right
  Diaconal: acceptable (?)
  Social: heathen, ethnic, polygamous, and all terms under the following
    subordinate categories:
  Lingual: famous, private, historical, representative
  Juridical: customary (or Social? or Technical?), traditional (?),
    independent (?), colonial, royal, national
  Economic: unmet, own, precious, rich, wealthier
  Laboratorial:
    Cognitional: paramount, special, ultimate, simple, profound
    Technical: pastoral, difficult
    Aesthetic: magnificent, pleasant
Behavioral: humble, lazy, shy, curious, aloof, proud
Biotic: young, elderly, flowery
Physical:
  Energetic: clean (or Aesthetic?), substantial, corrugated, fenceless, roast,
    hard, bitter, fragile, hot, inland, stony, coastal, spotless, dark, orange.
    green, blue, white, black, brown
  Kinematic: steady, earlier, old, new, ancient
  Mathematical:
    Spatial: highest, deep, huge, wide, northern, upper, distant, colossal,
      open, plump, small, thin, high, great
    Quantitative: thousand, two, 23 million, 10,000, 100,000, one, 500,
      uncounted, second, ten, three, 112, fewer, dozen, more, 13,000,
      3,900,000, four, many, double, much
    Aggregative: all, every, other

(The Mode or Function is followed by adjectives Weighted in it.)

A fairly good separation of adjectives into (1) Personal, (2) Behavioral, (3) Biotic, and (4) Physical obtains if one asks whether the adjective in question applies nonmetaphorically (1) only to men or what has been formed or evaluated ("magnificent," "pleasant") by men, or whether the adjective could apply to at least some (2) animals, (3) plants, or (4) Inorganic Creatures not greatly altered by men. Beyond these fairly obvious divisions, further division or order is much less visible.

Table 14

Mathematical Subfunctions

Views of the Mathematical Function	corresponding Subfunction	study	examples
Particle	Aggregative	Set Theory	Ac include Pa be included
Wave	Quantitative	Arithmetic	Ac multiply Pa number
Field	Spatial	Geometry	Ac extend Pa be bounded

Table 15

Dooyeweerd's aspects classify verbs

pistical: pray
ethical:
juridical: serve, obey, command (linguistic?)
aesthetic: surprise, expect (analytical?), hope
economic: render, inherit, reserve (biotic? historical?)
social: bow (juridical?), wear (historical?), adopt, attend, welcome, greet
linguistic: inquire, relate, tell, ask, describe, report, say, speak
historical: strap (?), make, bind (?), divide (?), build, send, weigh (?)
analytical: remember, know, think
psychical: wait (social?), yell, follow, lead, try (historical?), discover (ana-
    lytical?), leave, find (analytical?), sit, watch, wander, sense, look, arise,
    occupy (physical?), nuzzle, bask, retreat, stoop, return (kinematic?),
    kill, plod, see, lose, drink, nod, kneel, balance
biotic: die, live
physical: emit, ring, break, push, reflect, shake (kinematic?)
kinematic: change, stop, stretch, plunge, emerge, enter, come, endure, fall,
    swell, drift, climb, spread, disperse, stand (spatial?), crumple, go, join,
    smooth
spatial: crisscross, cover, encompass, include (but all these verbs can be taken
    in an active sense and put under kinematic)
numerical:

Table 16

Modes and Functions classify verbs

Personal:
  Sabbatical:
    Dogmatical: pray
    Presbyterial:
    Diaconal:
  Social:
    Lingual: command (Juridical?), greet (Economic?), inquire, relate,
      tell, ask, describe, report, speak, say
    Juridical: serve, obey, send
  Economic: render, inherit, reserve, bow, attend, welcome
  Laboratorial:
    Cognitional: remember, know, think
    Technical: adopt, strap, make, bind, divide, build, weigh
    Aesthetic: surprise, hope, expect, wear
Behavioral: see, look, sense, watch; yell, break, kneel, nod, drink, plod, kill,
  return, stoop, try, bask, nuzzle, arise, wander, sit, leave; find, lose, discover,
  lead, follow, retreat, occupy
Biotic: die, live
Physical:
  Energetic: stop, balance, shake, ring, smooth, emit, push, reflect, change,
    spread, disperse, stretch, crumple, join
  Kinematic: emerge, enter, come, endure, fall, drift, swell, climb, go,
    crisscross, cover, encompass, plunge (intransitive)
  Mathematical:
    Spatial: stand
  Numerical:
  Aggregative: include

---

by relabeling and reordering.[8] He has confessional, ethical, juridical,
economical, social, analytical, lingual, aesthetical, technical, psychical, bio-organic, physical, mathematical movement, spatial, numerical,
in that order. But the adjectives would presumably be split up among
aspects much as they are in Table 12. The possible reasons for
preferring Seerveld's order are not clear (see 3.131).

A similar classification can be done with verbs, but some of the
decisions of classification are as difficult as for the adjectives. All
verbs are interpreted as transitive if possible. See Tables 15 and 16.

NOTES TO APPENDIX 3

1. John M. Frame, *The Amsterdam Philosophy: A Preliminary Critique* (Phillipsburg, N. J.: Harmony Press, c. 1972), pp. 20-21, n. 3.

2. Hendrik Stoker, *Beginsels en metodes in die wetenskap* (Potchefstroom: Pro-Rege-Pers, 1961), p. 165; Calvin Seerveld, *A Christian Critique of Literature* . . . , Christian Perspectives Series 1964 (Hamilton, Ontario: Association for Reformed Scientific Studies, c. 1964), p. 33. The original aspect system is that of Herman Dooyeweerd, *A New Critique of Theoretical Thought* (Philadelphia: Presbyterian and Reformed, 1969), II.

3. "Here are meant the fundamental universal modalities of temporal being which do not refer to the concrete "what" of things or events, but are only the different modes of the universal "how" which determine the aspects of our theoretical view of reality" (*ibid.*, I, p. 3, n. 1). "This [qualifying] adjective denotes another modal aspect which, by means of an analogical moment of its structure, reveals its intermodal coherence with the original modus" (*ibid.*, II, pp. 67-68).

4. Cf. the discussion of "analogical use" in *ibid.*, II, pp. 61-72.

5. Joseph Judge, "The Zulus: Black Nation in a land of Apartheid," *National Geographic* 140 (December, 1971), pp. 738-775. The adjectives and verbs come from pp. 738-746.

6. Hendrik Stoker, *Beginsels*, p. 165.

7. *Ibid.*, pp. 164, 166.

8. Seerveld, *Critique*, p. 33.

11. Appendix 4

ON USING THE LAW OF CONTRADICTION

Contradiction ought to be avoided in argument. But what this implies depends on how one interprets the word 'contradiction' and on what kind of appeal to the "law of contradiction" one accepts as legitimate.

Let 'contradiction$_1$' stand for the word 'contradiction' as it is used loosely in everyday speech. Then the phrase 'law of contradiction' could be used with reference to any of the following.

law of contradiction$_1$: a contradiction$_1$ is not allowed.
law of contradiction$_2$: a statement cannot be both true and false at the same time and in the same way.
law of contradiction$_3$: "An object x cannot be both y and not-y."[1]
law of contradiction$_4$: some language is meaningful. It says something without saying everything, and also excludes some possibilities. A person who uses language ought to grant that his language is meaningful in this way.[2]
law of contradiction$_5$: no violations of Aristotle's logic are correct.[3]
law of contradiction$_6$: contradictions$_3$ obtained by syllogistic argument show that one premise is false.

And so we could go on. I take it that when a person speaks of using the law of contradiction$_i$ he means that an argument or statement or declaration which is shown to violate the law of contradiction$_i$ is thereby refuted (except that the law of contradiction$_6$ itself explains its use).

Now what results can be obtained by using the law of contradiction$_i$? The results depend on the value of i. Let us begin with one

199

of the more innocuous of the six, namely the law of contradiction$_2$. I agree with the law of contradiction$_2$, but I do not see how much can be obtained from it. For example, I do not see that it is sufficient to refute skepticism. Since the refutation of skepticism is one of the easiest of all refutations, it is a stringent test of the weakness of the law of contradiction$_2$.

Let us consider Gordon Clark's refutation of skepticism:

> Skepticism is the position that nothing can be demonstrated. And how, we ask, can you demonstrate that nothing can be demonstrated? The skeptic asserts that nothing can be known. In his haste he said that truth was impossible. And is it true that truth is impossible? For, if no proposition is true, then at least one proposition is true—the proposition, namely, that no proposition is true. If truth is impossible, therefore, it follows that we have already attained it.[4]

First of all, I agree that this argument is an adequate refutation of skepticism. But that is because I believe that it is legitimate to bring to bear all kinds of truths (about Scripture, God, Creation, etc.) in order to refute a falsehood.

Second, I grant that the skeptic is likely to get himself into all manner of difficulties if he attempts to answer Clark's own argument. For in the answer he will commit himself to much more truth. So let us say that we have a skeptic Joe who will say only, "Nothing can be demonstrated." Those are all the words that we can get out of him.

In a sense, then, anyone can "win the debate" with Joe simply because Joe will not reply to objections. Presumably a "refutation" claims to be more than this. Let us suppose, then, that Clark's argument wants to claim that Joe's statement involves a contradiction$_2$, and that the law of contradiction$_2$ alone suffices to refute it.

Now I propose to examine just how Clark's argument goes beyond the law of contradiction$_2$. It appeals to other things besides the fact that "a statement cannot be both true and false at the same time and in the same way"—or else it is involved in qualifications about what is meant by 'true,' 'false,' and 'same.'

a. Clark begins his argument with the following: "And how, we ask, can you demonstrate that nothing can be demonstrated?" This assumes too much. Joe has not demonstrated his statement, nor does he claim to demonstrate it. Rather, he says it.

b. "The skeptic asserts that nothing can be known." This cannot be obtained from Joe's statement unless (1) 'known' is being used as a synonym for 'demonstrated' (in ordinary English the two are not exactly synonymous)[5] and (2) it follows from "nothing can be demonstrated" that "Joe asserts that nothing can be demonstrated."

c. "In his haste he said that truth was impossible." How could this be known except by someone's *hearing* what Joe said? And from Clark's point of view, "sense experience" is epistemologically problematical. Moreover, once again not-strictly-synonymous terms are substituted (roughly, 'truth' for 'demonstrated,' 'impossible" for 'nothing can'). But, for the sake of argument, let us suppose that Joe commits himself to saying, "Truth is impossible."

d. "And it is true that truth is impossible?" The implied affirmation in this rhetorical question, namely, "It is true that truth is impossible," can be derived from "truth is impossible" only by a rather sophisticated decision about the proper use of the word 'true.' It is difficult to see how the law of contradiction$_2$ can be applied, in the form stated above, since it has not been established that any given statement is true, and neither has it been established that any given statement is false.

e. "For if no proposition is true, then at least one proposition is true—the proposition, namely, that no proposition is true." Here it has been assumed that 'no proposition is true' follows from 'truth is impossible' (or perhaps even from the earlier 'nothing can be demonstrated' which is supposed to be synonymous (?) with 'truth is impossible'). Does it then follow by the law of contradiction$_2$? But to apply the law we would have first to have a statement that is true, or that is false, and the statement that "no proposition is true" has not been shown to be either true or false. Joe has uttered it or asserted it (or something like it). But suppose we grant that 'no proposition is true' is true. How do we know that "at least one proposition is true"?

Because, presumably, "a statement cannot be both true and false at the same time and in the same way"? But who said otherwise? The law of contradition$_2$ in the present form *still* does not give anything, because still no one has yet said that any of our statements are false. We need an intermediate step, namely, a statement that if a is P, then at least one x is P. How do we get this?

Presumably what Clark really wants is to derive the statement "truth is possible." (To do this he has to go from using 'true' to using 'truth,' but I let that pass.) But we still have no false statement. Hence what we want is " 'no proposition is true' is false." (And the introduction of the word 'false' cannot be accomplished without some kind of assumptions on the relation of the word 'not' or the prefix 'im-' to the word 'false,' and probably an introduction of the law of excluded middle in some form or other.)

When we have done all this, then we must show that 'no proposition is true' has the same sense in both cases of its use. And this will beg the question with the skeptic, since the point of skepticism may well be that the two words 'true' in ' 'no proposition is true' is true' have different senses.

Next, let us try the law of contradiction$_1$. Whether a Christian would agree with it depends on who is using the word 'contradiction$_1$.' Many non-Christians would say that the doctrine of the Trinity involves a contradiction$_1$. They would say that I Samuel 15:29 and I Samuel 15:11, 35 involve a contradiction$_1$. Clark and other Christians would say that there is no contradiction$_1$. This difference of usage is presumably what Van Til has in mind in his objections against a "neutral" law of contradiction.

Let us then try the law of contradiction$_3$. Christian teaching says that Jesus Christ in the flesh is man and non-man (God), omniscient and nonomniscient, finite and nonfinite.[6] Hence, Christianity violates the law of contradiction$_3$ unless we qualify it by saying (as presumably Clark would do) that these two sides are not true in the same sense and in the same way.

Is it then the case that 'man' or 'finite' is being used in two different senses? Presumably that is not it. The Christian might want to say that 'nonfinite' applies to the divine nature of Christ and 'finite' to the human nature of Christ. Very well.[7] But the non-Christian would certainly object that this is saying little more than that the word 'finite' applies to the finite and the word 'nonfinite' applies to the nonfinite regarding Christ. If this method of evading contradiction₃ can be used here, then the non-Christian will proceed to claim that different "senses" or "ways" are involved every time he is caught in a supposed "contradiction₃." Hence spelling out the implications of 'sense' or 'way' virtually involves an acceptance of or repudiation of the Christian faith.

Next, the law of contradiction₄. I agree with it, but I do not think that it will get us far in an argument, unless perhaps the meaning of 'meaningful' is made more precise.

What about the law of contradiction₅? Since Aristotle's logic includes the law of contradiction₃, it is beset with the same problems as this law.

Next the law of contradiction₆. This law is in any case beset with the problems of the law of contradiction₃ But there is an additional problem. As is well known, this law as it stands does not always work properly when applied to ordinary natural languages. This is so, not only because of the vagueness of terms in natural languages, but because of the following phenomenon.

P1   Joe has stopped beating his wife.
P2   Joe has not stopped beating his wife.

Presumably P1 and P2 together are a violation of the law of contradiction₃. Hence either P1 is false or P2 is false.

So people using natural language will conclude that Joe at some time or other has been beating his wife. The problem can be avoided in at least two ways. (1) We may employ a technical term 'false₂,' which applies to any statement to which 'true' does not apply (a statement is false₂ if and only if we would not want to say, "This statement is true"). Or (2), we may insist on "purifying" or altering natural language before putting it into a syllogism. These methods, applied rigorously, will result in the syllogistic method applying workably to a situation so constructed that it must apply.

This procedure may no doubt be useful. But it does not solve all our problems. We must still decide, in a given case, whether natural language is close enough to the artificial language so constructed to warrant our carrying over the conclusions to everyday life. Hence, no one of the above senses of the "law of contradiction" provides a suitable "neutral" test of religious truth. To all this, I suppose Gordon Clark would reply that by "the law of contradiction" or "the laws of logic" he means "the thing itself" rather than fallible human attempts to apply it. In that case, I would not quarrel with him. I do not see how he differs from Van Til.

But perhaps we might press Clark further. In Clark's view, "the law of contradiction" (call it law of contradiction$_7$) is a truth or a number of truths or at any rate something in God's mind. Not only that, but it appears that the law of contradiction$_7$ would have to be *defined* in terms of God, because human notions of the law of contradiction can be either too vague or in part incorrect or both (think especially of contradiction$_1$). The law of contradiction$_7$ thus depends on God for its meaningfulness or reality or value. I suppose also that a *denial* of the law of contradiction$_7$ is itself a contradiction$_7$.

Now would Clark agree that a denial of the existence of God implies a denial of the law of contradiction$_7$? Since the law of contradiction$_7$ depends on God, presumably this is so. Hence a denial of the existence of God is a contradiction$_7$. Hence God exists. Thus Clark actually has available a theistic proof$_7$!

Of course, this proof$_7$ is simply another version of the "presuppositional" apologetic of Cornelius Van Til. If the above interpretation is correct (and it may not be), Clark has simply concealed his presuppositionalism under the term 'contradiction.'

NOTES TO APPENDIX 4

1. From Gordon H. Clark, *A Christian View of Men and Things* (Philadelphia: Presbyterian and Reformed, 1952), p. 292.
2. See *ibid.* and Ronald H. Nash, ed., *The Philosophy of Gordon H. Clark* (Philadelphia: Presbyterian and Reformed, 1968), p. 131, for something near to this.
3. Still further versions of the law of contradiction could be obtained by substituting the names of other logicians for 'Aristotle.'

4. Clark, *Christian View,* p. 30.

5. 'God can be known' is certainly true, and would be granted by the average Christian, while 'God can be demonstrated' might be challenged by people who do not accept theistic proofs.

6. I assume the reader will see that to say that Jesus Christ is nonomniscient is not the same as to say that Jesus Christ is not omniscient or to say that it is not true that Jesus Christ is omniscient. The first of these three is true and the latter two are false. At least this is true of my idiolect. If someone else's idiolect differs from mine, I think I could adapt myself for the purpose of communication.

7. Though Nestorianism is a danger if this is all that we say.

GLOSSARY

Section numbers enclosed in parentheses refer to places where the term in question is discussed but a formal description is not given.

Accomplishment. 3.21, 3.26 (3.22).

The Accomplishment Period of redemption is the period comprising Jesus' birth, life, death, resurrection, and ascension—the period covered by the Gospels. The Adamic Accomplishment Period is the period of Adam's probation (Gen. 2:4–3:7), from the creation of man to his fall.

(In contrast with Preparation and Application.)

actional Functions. (3.1213)

The actional Functions are the Active, Middle, and Passive Functions; that is, those three subdivisions within the Personal Mode obtained by focusing on the degree of initiative of personal actors.

(In contrast with ordinantial Functions and official Functions.)

Active. 3.1213.

The Active Function is that part of the Personal Mode having to do with activities and characteristics where the persons in question take some kind of initiating role, where they are giving, where they are affected, as it were, from inside out.

(In contrast with Middle and Passive Functions.)

Adamic. 3.26.

'Adamic' is an adjective used to construct Periods analogous to Periods already described with reference to the work of Christ. Adamic Periods are those Periods in the history of those "in" Adam, analogous to the Periods in the history of those in Christ. Thus the Adamic Preparation Period (Gen. 1:1–2:3) is the

Period preparing for Adam, just as the Preparation Period is the period preparing for Christ. Similarly for the Adamic Accomlishment Period (Gen. 2:4–3:7) and the Adamic Application Period (from Gen. 3:8 to the end of time).

Administrative. 3.321 (3.322).

The Administrative aspect of a relation is that part of the relation having to do with the Kingly Function. In particular, the Administrative aspect of the Covenantal Bond is the Covenantal administration.

(In contrast with Locutionary and Sanctional aspects.)

Adumbrative. 3.122.

The Adumbrative Prophetic, Kingly, Priestly, Active, Middle, and Passive Functions are the adumbrative forms of these Functions found within the non-Personal Modes.

Aesthetic. 3.1212.

The Aesthetic Function is that part of the Personal Mode covered by the Priestly and Laboratorial Functions; that is, the Aesthetic Function consists of the bundle of characteristics associated *both* with priestly sharing and communion *and* with man's labor.

(In contrast with Cognitional, Technical, Diaconal, and Economic Functions.)

Aesthetics. 3.1212.

Aesthetics is the study of the Aesthetic Function.

(In contrast with Logic, Technology, Diaconology, and Economics.)

Aggregative. 10.

The Aggregative Subfunction is that part of the Mathematical Function that the Particle View focuses on. That is, it consists of the closure properties and features found in the "meaning" side of the Physical Mode.

(In contrast with the Quantitative and Spatial Subfunctions.)

Angelology. 2.41.

Angelology is the Study of Angels.

Angels. 2.41.

Angels are personal Creatures who belonged to Heaven at the time when they were created.

Animal Kingdom. 2.432 (3.11).

The Animal Kingdom is one of the three divisions of the Subhuman Kingdom laid out for man in Genesis 1:28-30. It consists of animals, characterized as moving and breathing.

(In contrast with Plant and Inorganic Kingdoms.)

Anthropology. 2.4 (3.11).

Anthropology is the study of Men.

(In contrast with Ouranology and Cosmology.)

Application. 3.21, 3.26 (3.23).

The Application Period of redemption is the period of the application of the benefits that Christ has won by his work—the period covered by the Book of Acts and onwards. The Adamic Application Period is the period after the fall of Adam, from Genesis 3:8 onward to the end of time.

(In contrast with Preparation and Accomplishment Periods.)

Appraisive. 3.27.

The Appraisive aspect of events is that aspect involving the Priestly Function. Sometimes there can be a temporal separation between an Appraisive phase and Vocative and Dynamic phases.

(In contrast with Vocative and Dynamic.)

Aquatic Kingdom. 2.431 (2.432).

The Aquatic Kingdom is that part of the Cosmos consisting of the waters (seas, rivers, etc.) and their inhabitants.

(In contrast with the Terrestrial Kingdom and Heaven.)

Arithmetic. 10.

Arithmetic is the study of the Quantitative Subfunction.

(In contrast with the Set Theory and Geometry.)

Axial. 3.332, 3.321, 3.322 (3.323, 3.3243).

Axial Views are views focusing on the relations among parties, when the parties are within some relationship. In particular, the Axial View of the Covenantal Bond focuses on the Covenantal Bond itself.

(In contrast with Polar Views.)

Axiological. (6.121).

Axiological means pertaining to value. Axiological Study is Study of value.

(In contrast with Ontological and methodological Study.)

Axiology. 4.2 (3.35).

Axiology is the study of value.

(In contrast with ontology and methodology.)

Basilic. 3.1212.

See Kingly.

Basilics. 3.1212.

Basilics is the study of the Kingly Function.

(In contrast with Prophetics and Hieratics.)

Behavioral. 3.11.

The Behavioral Mode is the bundle of characteristics that the Animal Kingdom has in addition to those of the Plant and Inorganic Kingdoms.

(In contrast with Personal, Biotic, and Physical Modes.)

Behaviorology. 3.11.

Behaviorology is the study of the Behavioral Mode.

(In contrast with Ethology, Biology, and Physics.)

Beneficence. 6.12.

Beneficence is Personal activity with Priestly Weight, or the result of such activity.

(In contrast with Study and Technics.)

Biology. 3.11.

Biology is the study of the Biotic Mode.

(In contrast with Ethology, Behaviorology, and Physics.)

Biotic. 3.11.

The Biotic Mode is the bundle of characteristics that the Plant

Kingdom has in addition to those of the Inorganic Kingdom. (In contrast with Personal, Behavioral, and Physical Modes.)

Bond. 3.322, 3.321.

The Bond is to Dominical Bond (the totality of God's relations to himself and to Creation) or the Covenantal Bond (that part of the Dominical Bond which is the pact under sanctions, revealed in Scripture, between God and his people), or a Servient Bond (that part of the Covenantal Bond that pertains to a given Creature).

(In contrast with God and Creation.)

Botany. 2.432.

Botany is the study of the Plant Kingdom.

(In contrast with Zoology and Inorganics.)

Boundary. 6.122.

Boundary Study is Study that concentrates on questions of a "boundary" character, that is, questions that, in a given temporal stage of history, cannot reach definitive resolution by Cosmic Men.

(In contrast with Special Study.)

Canonical. 6.123.

Canonical Study is the Covenantal Word of God, or the act of its production.

(In contrast with Evangelical and Speculative Study.)

Christology. 2.2.

Christology is the study of the Son of God.

(In contrast with Patrology and Pneumatology.)

church. (3.331).

The church is the people of God, those in union with Christ, especially in the Application Period.

(In contrast with the Human Kingdom.)

Cognitional. 3.1212.

The Cognitional Function is that part of the Personal Mode covered by the Prophetic and Laboratorial Functions; that is, the Cognitional Function consists of the bundle of characteristics associated *both* with prophetic communication *and* with man's labor.

(In contrast with Technical, Aesthetic, Dogmatical, and Lingual Functions.)

Contrast. 5.21.

The Contrast of an Item involves those features that identify it and contrast it with other Items. "Items which are independently, consistently different are in contrast."

(In contrast with Variation and Distribution of Items.)

Corporate. 3.23, 3.26.

Corporate Periods are decisive stages with respect to groups of men (in particular, the church). The Corporate Generational, Developmental, and Culminational Application Periods are, respectively, (1) the time of Pentecost and the founding of the church (Acts), (2) the history of the church from its founding to the coming of Christ (Revelation), and (3) the glorification of the church at the Coming of Christ, and the time to follow. Likewise with Corporate Adamic Periods.

(In contrast with Individual Periods.)

Cosmic Human Kingdom. 2.42.

The Cosmic Human Kingdom is that part of the Human Kingdom in the Cosmos. Cosmic Men are Men in the Cosmos.

(In contrast with the Heavenly Human Kingdom.)

Cosmology. 2.4.

Cosmology is the study of the Cosmos.

(In contrast with Ouranology and Anthropology.)

cosmonomic philosophy. (8.2, 8.3, 9., 10.).

Cosmonomic philosophy is the school of philosophy within the Reformed community. Building on the work of Abraham Kuyper, a number of men—chiefly Herman Dooyeweerd, Dirk H. Th. Vollenhoven, and Hendrik G. Stoker—founded this school. Generally speaking, it is characterized by (a) emphasis on the comprehensive character of God's law, (b) a naïve/ theoretical distinction, (c) interest in the nature and relationship of the sciences, (d) interest in the influence of religious motives on the historical development of thought, and (e) "theoretical" analysis of the "cosmos" into "modal aspects" or "spheres," each of which is "sovereign" and "irreducible."

Cosmos. 2.4, 2.42.

The Cosmos is that part of Creation which is not Heaven; or, equivalently, which includes men made of dust and subhuman Creation to be ruled by such men.

(In contrast with Heaven and the Human Kingdom.)

Covenantal. 3.321, 3.332 (3.322, 3.32432).

The Covenantal Bond is the pact under sanctions, revealed in Scripture, between God and his people. Scripture sums up the Covenantal Bond in the words, "I will be your God, and you shall be my people." The Covenantal Bond includes both law, administration, and sanctions of covenants.

(In contrast with Dominical and Servient Bond.)

The Covenantal View of a structure is the view from the standpoint of the Covenantal Bond.

(In contrast with Dominical and Servient Views.)

covenant-breaker. 5.33.

A covenant-breaker is a Man with a fundamental (Sapiential) orientation of rebelling against God. He is not a member of God's kingdom.

(In contrast with covenant-keepers.)

covenant-keeper. 5.33.

A covenant-breaker is a Man with a fundamental (Sapiential) orientation toward serving God. He is a member of God's kingdom.

(In contrast with covenant-breakers.)

Creation. 2.1.

Creation is everything that has been created by God, i.e., everything that has a beginning.

(In contrast with God. *See* Mediator.)

Creator. (2.2).

The Creator is God.

(In contrast with Creation. *See* Mediator.)

Creature. 2.1.

A Creature is a thing in Creation.

(In contrast with God. *See* Mediator.)

Culminational. 3.22, 3.23, 3.24, 3.25, 3.26.

A Culminational Period is a final period of summing up, in a sequence of three periods. The Priestly Function is in greater prominence in a Culminational Period. The Culminational Accomplishment Period is from the trial of Jesus to his resurrection, ascension, and enthronement; an Individual Culminational Application Period is the period of glory in a Christian's life initiated by his death or by the return of Christ; the Corporate Culminational Application Period consists of the glorification of the church at the coming of Christ, and the time to follow; the Culminational Preparation Period is the OT period of execution of sanctions (roughly I Kings 8 or 11 to Neh.).

(In contrast with Generational and Developmental Periods.)

The Culminational View is the view of an event or complex of events which sees the events primarily in terms of fulfilling the past behind the events in question. (In contrast with Generational and Developmental Views.)

description. (2.0).

A description is a specification of how a technical (capitalized) term will be used.

Developmental. 3.22, 3.23, 3.24, 3.25, 3.26.

A Developmental Period is a middle period of development, in a sequence of three periods. The Kingly Function is in greater prominence in a Developmental Period. The Developmental Accomplishment Period is from Jesus' temptation in the wilderness to Gethsemane; an Individual Developmental Application Period is the time of a Christian's walking with Christ in this world; the Corporate Developmental Application Period is the period from the founding of the church in Acts to the coming of Christ; the Developmental Preparation Period is the OT period of establishment of kingship and kingdom (roughly Judges–I Kings 7 or 10).

(In contrast with Generational and Culminational Periods.)

Diaconal. 3.1212.

The Diaconal Function is that part of the Personal Mode covered by the Priestly and Sabbatical Functions; that is, the Diaconal Function consists of the bundle of characteristics associated *both* with priestly sharing and communion *and* with sabbatical worship.

(In contrast with Dogmatical, Presbyterial, Economic, and Aesthetic Functions.)

Diaconology. 3.1212.

Diaconology is the study of the Diaconal Function.

(In contrast with Dogmatics, Presbyteriology, Economics, and Aesthetics.)

Differentiated. 3.333.

A Differentiated Unit is a Unit with rather clearly discernible Weight in one unique Function of the nine Functions Dogmatic Presbyterial, Diaconal, Lingual, Juridical, Economic, Cognitional, Technical, or Aesthetic.

(In contrast with Semidifferentiated and Undifferentiated Units.)

Distribution. 5.21.

The Distribution of an Item is comprised by the neighborhoods in which it may occur. This may be further analyzed into Distribution in class (Particle), in sequence or location (Wave), and in system (Field).

(In contrast with Contrast and Variation.)

Dogmatical. 3.1212.

The Dogmatical Function is that part of the Personal Mode covered by the Prophetic and Sabbatical Functions; that is, the Dogmatical Function consists of the bundle of characteristics associated *both* with prophetic meaning and communication *and* with sabbatical worship.

(In contrast with Presbyterial, Diaconal, Lingual, and Cognitional Functions.)

Dogmatics. 3.1212.

Dogmatics is the study of the Dogmatical Function.

(In contrast with Presbyteriology, Diaconology, Linguistics, and Logic.)

Dominical. 3.322, 3.332, (3.32431).

The Dominical Bond is the totality of God's relations to himself and to Creation. Thus it includes the Covenantal Bond.

The Dominical View of a structure is the view from the standpoint of the Dominical Bond.

(In contrast with Covenantal and Servient Bond.)

Dynamic. 3.27.

The Dynamic aspect of events is that aspect involving the Kingly Function. Sometimes there can be a temporal separation between a Dynamic phase and Vocative and Appraisive phases.

(In contrast with Vocative and Appraisive.)

Economic. 3.1212

The Economic Function is that part of the Personal Mode covered by the Priestly and Social Functions; that is, the Economic Function consists of the bundle of characteristics associated *both* with priestly sharing and communion *and* social activity.

(In contrast with Lingual, Juridical, Diaconal, and Aesthetic Functions.)

Economics. 3.1212.

Economics is the study of the Economic Function.

(In contrast with Linguistic, Jurisprudence, Diaconology, and Aesthetics.)

Emphasizing Reductionism. 3.133.

Emphasizing Reductionism is preference for or preoccupation with one (or a small number of) viewpoint(s), Mode(s), Function(s), or other Item(s) when one attempts to discuss and interpret some subject-matter.

(In contrast with Exclusive and Slippery Reductionism.)

Energeticology. 3.1213.

Energeticology is the study of the Active Function.

(In contrast with Mesology and Patheticology.)

Energetics. 3.122.

Energetics is the study of the Adumbrative Priestly Function of the Physical Mode.

(In contrast with Mathematics and Kinematics.)

Ergology. 3.1211.

Ergology is the study of the Laboratorial Function.

(In contrast with Liturgiology and Sociology.)

Ethics. 4.2 (4.1).

Ethics is Sanctional Anthropological Axiology, that is, the study of what human persons, deeds, intentions, and dispositions warrant approval.

(In contrast with Locutionary Anthropological Axiology, Administrative Anthropological Axiology, Sanctional Ouranological Axiology, and Sanctional Theriological Axiology.)

Ethology. 3.11.

Ethology is the study of the Personal Mode.

(In contrast with Behaviorology, Biology, and Physics.)

Evangelical. 6.123.

Evangelical Study is Study of the Covenantal Word of God, with the purpose of communicating what this Word says.

(In contrast with Canonical and Speculative Study.)

Exclusive Reductionism. 3.133.

Exclusive Reductionism is the insistence on the exclusive correctness of one's own form of Emphasizing Reductionism; that is, insistence on the innate superiority of one's own special vocabulary.

(In contrast with Emphasizing Reductionism and Slippery Reductionism.)

Exegesis. 6.123.

Exegesis is Evangelical Study that answers the question, "What does this [some particular] passage of the Bible say?" (In contrast with Systematic Theology and biblical theology.)

Existential. 4.1, 4.2 (3.35).

The Existential Perspective on Axiology is that way of looking at the value of Items which focuses on the Items themselves in their dynamic development. In particular, the Existential Perspective on Ethics focuses on persons and their motives. (In contrast with Normative and Situational Perspectives.)

Exploration. (6.122).

Exploration is Boundary Technics. It is Technics that concentrates on tasks of a "boundary" character, that is, tasks that, in a given temporal stage of history, cannot be performed by Cosmic Men on the basis of already-agreed-upon methodology and justification. (In contrast with Special Technics.)

Field. 3.123.

A Field View is a view focusing on the interdependent characteristics and relations of Items. (In contrast with Particle and Wave Views.)

First Polar View. 3.321, 3.322 (3.332, 3.323, 3.3243).

The First Polar View of the Bond focuses on what God himself does with reference to the Bond. (In contrast with Axial and Second Polar Views.)

Functions. (3.12).

Functions are subdivisions within or parts of Modes. A complete list of Functions is as follows: Sabbatical, Social, Laboratorial; Prophetic, Kingly, Priestly; Dogmatical, Presbyterial, Diaconal, Lingual, Juridical, Economic, Cognitional, Technical, Aesthetic; Active, Middle, Passive; Adumbrative Prophetic, Adumbrative Kingly, Adumbrative Priestly, Adumbrative Active, Adumbrative Middle, Adumbrative Passive. (In contrast with Modes and Subfunctions.)

Generational. 3.22, 3.23, 3.24, 3.25, 3.26.

A Generational Period is a first period of initiation, in a sequence of three periods. The Prophetic Function is in greater prominence in a Generational Period. The Generational Accomplishment Period is the period comprising the birth narratives of Jesus; an Individual Generational Application Period of the Christian's redemption is his time of conversion, of initiation into God's people; the Corporate Generational Application Period is the time of Pentecost and the founding of the church (Acts); the Generational Preparation Period is the Mosaic period of the OT (roughly Gen. 3:8–Joshua).

(In contrast with Developmental and Culminational Periods.)

The Generational View is the view of an event or complex of events which sees the events primaritly in terms of presaging a future. (In contrast with Developmental and Culminational Views.)

Genuine. 6.120.

Genuine Study, or Technics, or Beneficence, is such Study, Technics, or Beneficence *for* God and his kingdom.

(In contrast with Pseudo Study.)

Geography. 2.431.

Geography is the study of the Terrestrial Kingdom.

(In contrast with Oceanography and Ouranology.)

Geometry. 10.

Geometry is the study of the Spatial Subfunction.

(In contrast with Set Theory and Arithmetic.)

God. (1.1), (2.1).

God is Yahweh, the God who has told us about himself in the Bible, which is his word.

(In contrast with idols and with Creation. *See* Mediator.)

Heaven. 2.40 (2.41).

Heaven is that part of Creation not accessible to the ruling powers of men made of dust.

(In contrast with the Cosmos and the Human Kingdom.)

Heavenly Human Kingdom. 2.42.

    The Heavenly Human Kingdom is that part of the Human Kingdom in Heaven. Heavenly Men are Men in Heaven.

    (In contrast with the Cosmic Human Kingdom.)

Hieratic. 3.1212.

    See Priestly.

Hieratics. 3.1212.

    Hieratics is the study of the Priestly Function.

    (In contrast with Prophetics and Basilics.)

History. 6.121.

    History is Refined Study by Cosmic Men, of the (temporal) past of the Human Kingdom, particularly the Technical past.

    (In contrast with Study of the present and the future, or Modal or Structural Study.)

Human Kingdom. 2.42.

    The Human Kingdom is all Men taken together.

    (In contrast with Heaven and the Subhuman Kingdom.)

Humanities. 6.121.

    Humanities is Refined Modal Study by Cosmic Men, of the Personal Mode and various Functions within it, especially when such Study has methodological dissimilarity to Natural Science.

    (In contrast with Social Science and Natural Science.)

Incarnate Christology. 2.3.

    Incarnate Christology is the study of the man Christ Jesus (I Tim. 2:5).

    (In contrast with Theology Proper and Ktismatology.)

Individual. 3.23.

    Individual Periods are decisive stages with respect to individual men. The Individual Generational, Developmental, and Culminational Application Periods of the Christian's redemption are, respectively, (1) the time of conversion, of initiation into God's people; (2) the period of walking with Christ in this world; and (3) the period of glory initiated by death or the return of Christ.

    (In contrast with Corporate Periods.)

Inorganic Kingdom. 2.432.

The Inorganic Kingdom is one of the three divisions of the Subhuman Kingdom laid out for man in Genesis 1:28-30. It consists of nonliving things.

(In contrast with the Animal and Plant Kingdoms.)

Inorganics. 2.432.

Inorganics is the study of the Inorganic Kingodm.

(In contrast with Zoology and Botany.)

Institution. 3.333.

An Institution is a Particulate Societal Unit; that is, it is a Unit, including Men in its internal substructure, that we regard normally as a unified whole enduring as more or less the same over a time span.

(In contrast with Societal Transactions and Societal Relationships.)

Interlocking. (3.132, 3.25, 3.34).

Interlocking of Items involves the overlap of Items, vagueness of boundary between the Items, mutual dependence of the Items, and inability to talk about, use, or appreciate one Item without indirectly involving the others.

(In contrast with order and luxuriance.)

Item. 3.123.

An Item is anything that Man selects for notice or study.

(In contrast with View and Man.)

Juridical. 3.1212.

The Juridical Function is that part of the Personal Mode covered by Kingly and Social Functions; that is, the Juridical Function consists of the bundle of characteristics associated *both* with kingly rule and power *and* with social activity.

In contrast with Lingual, Economic, Presbyterial, and Technical Functions.)

Jurisprudence. 3.1212.

Jurisprudence is the study of the Juridical Function.

(In contrast with Linguistics, Economics, Presbyteriology, and Technology.)

Kinematics. 3.122.

Kinematics is the study of the Adumbrative Kingly Function of the Physical Mode.

(In contrast with Mathematics and Energetics.)

Kingdom. 2.432, 2.431, 2.42, 2.4.

A Kingdom is a major ontological subdivision of Creation. If no qualifying adjective is present, "Kingdom" means any one of four Kingdoms: the Human Kingdom, the Animal Kingdom, the Plant Kingdom, or the Inorganic Kingdom. But the same word is used also with other qualifying adjectives: Subhuman Kingdom, Terrestrial Kingdom, and Aquatic Kingdom.

(In contrast with Creation and Creature.)

Kingly. 3.1212.

The Kingly or Basilic Function is that part of the Personal Mode having to do with activities, states, characteristics, etc., of a predominantly kingly sort; that is, the Kingly Function has to do with rule, power, mastery.

(In contrast with Prophetic and Priestly Functions.)

Ktismatology. 2.1.

Ktismatology is the study of Creation.

(In contrast with Theology Proper and Incarnate Christology.)

Laboratorial. 3.1211.

The Laboratorial Function consists of that part of the Personal Mode having to do with activities normally characteristic of the ordinance of labor, or equivalently, of a person's relation to the Theric Kingdom.

(In contrast with Sabbatical and Social Functions.)

Law. 3.3241.

The Law is the Covenantal Locution of God as king.

(In contrast with the Word of God and with particular commandments of God.)

Lingual. 3.1212.

The Lingual Function is that part of the Personal Mode covered by the Prophetic and Social Functions; that is, the Lingual Function consists of the bundle of characteristics associated *both* with prophetic meaning and comunication *and* with social activity.

(In contrast with Juridical, Economic, Dogmatical, and Cognitional Functions.)

Linguistics. 3.1212

Linguistics is the study of the Lingual Function.

(In contrast with Jurisprudence, Economics, Dogmatics, and Logic.)

Liturgiology. 3.1211.

Liturgiology is the study of the Sabbatical Function.

(In contrast with Sociology and Ergology.)

Locutionary. 3.321 (3.322).

The Locutionary aspect of a relation is that part of the relation having to do with the Prophetic Function. In particular, the Locutionary aspect of the Covenantal Bond is the Covenantal words.

(In contrast with Administrative and Sanctional aspects.)

Logic. 3.1212.

Logic is the study of the Cognitional Function.

(In contrast with Technology, Aesthetics, Dogmatics, and Linguistics.)

luxuriance. (3.133).

The luxuriance of Items is their distinctness from one another, the inability to merge one completely into the others, or to perform a valid Exclusive or Slippery Reductionism.

(In contrast with order and interlocking.)

Mathematics. 3.122.

Mathematics is the study of the Adumbrative Prophetic Function of the Physical Mode.

(In contrast with Kinematics and Energetics.)

Mediator. 2.3 (3.323).

The Mediator is "the man Christ Jesus" (I Tim. 2:5).

(*See* God and Creation.)

Men. 2.4.

Men are those Creatures in the image of God called to subdue the earth, have dominion over the animals, etc.

(In contrast with Heaven and the Subhuman Kingdom.)

Mesology. 3.1213.

Mesology is the study of the Middle Function.

(In contrast with Energeticology and Patheticology.)

methodology. (3.0, 3.35).

Methodology is anything that answers the question, "How do Items function?"

(In contrast with ontology and Axiology.)

Middle. 3.1213.

The Middle Function is that part of the Personal Mode having to do with a mutual interchange, a sharing.

(In contrast with Active and Passive Functions.)

Modal. 6.121.

Modal Study is Study whose subject-matter is (chiefly or focally) modality.

(In contrast with Temporal and Structural Study.)

modality. (3.0, 3.1, 3.35).

Modality is the more or less constant characteristics of Items, especially of Kingdoms, of Creatures, and of God.

(In contrast with temporality and structurality.)

Modes. 3.11.

A Mode is the bundle of characteristics that a Kingdom has in addition to those of the lower Kingdoms. The four Modes are Personal, Behavioral, Biotic, and Physical.

(In contrast with Functions and Subfunctions, and with Periods and structures.)

Natural Science. 6.121.
>Natural Science is Refined Modal Study by Cosmic Men, of the Behavioral, Biotic, and Physical Modes.
>(In contrast with Social Science and Humanities.)

Normative. 4.1, 4.2 (3.35).
>The Normative Perspective on Axiology is that way of looking at the value of Items which focuses on the rules concerning value (whether God's or Men's). In particular, the Normative Perspective on Ethics focuses on God's commands with respect to personal behavior.
>(In contrast with Existential and Situational Perspectives.)

Obligatory. 3.333.
>Obligatory Institutions are those mentioned explicitly in Scripture, which are such that, if a person belongs to the Institution in question, he ordinarily ought not to withdraw his participation except on dissolution of the Institution. The state, the family, marriage, and the church are Obligatory Institutions.
>(In contrast with Strategic and Voluntary Institutions.)

Oceanography. 2.431.
>Oceanography is the study of the Aquatic Kingdom.
>(In contrast with Geography and Ouranology.)

official Functions. (3.1212).
>The official Functions are the Prophetic, Kingly, and Priestly Functions; that is, those three subdivisions within the Personal Mode obtained by focusing on the offices of prophet, king, and priest.
>(In contrast with ordinantial Functions and actional Functions.)

Ontological Study. (6.121)
>Ontological Study is Study of what there is.
>(In contrast with methodological Study and Axiological Study.)

ontology. (2.), (3.35).
>Ontology is anything that answers the question, "What is there?"
>(In contrast with methodology and Axiology.)

order. (3.131).

An order of Items is a sequential arrangement according to some criterion of complexity, time, mutual relationship, or the like.

(In contrast with interlocking and luxuriance.)

ordinantial Functions. (3.1211).

The ordinantial Functions are the Sabbatical, Social, and Laboratorial Functions; that is, those three subdivisions within the Personal Mode obtained by focusing on the activity of Men with respect to the three "creation ordinances," or on the activity of men with respect to God, to the Human Kingdom, and to the Subhuman Kingdom.

(In contrast with official and actional Functions.)

Ouranology. 2.4.

Ouranology is the study of Heaven.

(In contrast with Anthropology and Cosmology.)

Particle. 3.123 (3.35).

A Particle View by a Man is a focusing on Items with their closure properties, including an ordering of Items in a taxonomy according to some set of features convenient for the purpose in hand.

(In contrast with Wave and Field Views.)

Particulate. 3.333.

A Particulate Unit is a structure that we regard normally as a unified whole enduring as more or less the same over a time span.

(In contrast with Undulatory and Relational Units.)

Passive. 3.1213.

The Passive Function is that part of the Personal Mode having to do with activities and characteristics where the persons in question take some kind of responding role, where they are receiving, where they are affected, as it were, from outside in.

(In contrast with Active and Middle Functions.)

Patheticology. 3.1213.
> Patheticology is the study of the Passive Function.
> (In contrast with Energeticology and Mesology.)

Patrology. 2.2.
> Patrology is the study of God the Father.
> (In contrast with Christology and Pneumatology.)

Period. 3.2.
> A Period is a temporal subdivision of the temporal, historical working out of God's plan with respect to himself and Creation. Each Period is a major discernible phase or stage of this working out. But some Periods are subdivisions of other Periods. A complete list of Periods occurring in this book is as follows: Preparation Period, Accomplishment Period, Application Period, Generational Accomplishment Period, Developmental Accomplishment Period, Culminational Accomplishment Period, Individual Generational Application Period, Individual Developmental Application Period, Individual Culminational Application Period, Corporate Generational Application Period, Corporate Developmental Application Period, Corporate Culminational Application Period, Generational Preparation Period, Developmental Preparation Period, Culminational Preparation Period, Adamic Preparation Period, Adamic Accomplishment Period, Adamic Application Period, Generational Adamic Preparation Period, Developmental Adamic Preparation Period, Culminational Adamic Preparation Period, Generational Adamic Accomplishment Period, Developmental Adamic Accomplishment Period, Culminational Adamic Accomplishment Period, Corporate Generational Adamic Application Period, Corporate Developmental Adamic Application Period, Corporate Culminational Adamic Application Period.
> (In contrast with Modes and structures.)

Personal. 3.11.
> The Personal Mode is the bundle of characteristics that the Human Kingdom has in addition to those of the Animal, Plant, and Inorganic Kingdoms.
> (In contrast with Behavioral, Biotic, and Physical Modes.)

Perspective. 4.1, 4.2 (3.35).
   A Perspective is one of three ways of looking at the value of
Items. The three are the Normative Perspective, the Existential
Perspective and the Situational Perspective.
   (*See* View.)

Philosophy. 6.122.
   Philosophy is Refined Boundary Study by Cosmic Men.
   (In contrast with Science and Theology.)

Physical. 3.11.
   The Physical Mode is the bundle of characteristics that the
Inorganic Kingdom has, and shares with other Kingdoms.
   (In contrast with Personal, Behavioral, and Biotic Modes.)

Physics. 3.11.
   Physics is the study of the Physical Mode.
   (In contrast with Ethology, Behaviorology, and Biology.)

Plant Kingdom. 2.432 (3.11).
   The Plant Kingdom is one of the three divisions of the Subhuman Kingdom laid out for man in Genesis 1:28-30. It consists of plants, characterized as green, growing, and reproducing.
   (In contrast with Animal and Inorganic Kingdoms.)

Pneumatology. 2.2.
   Pneumatology is the study of the Holy Spirit.
   (In contrast with Patrology and Christology.)

Polar. 3.332.
   Polar Views of a structure are views with a focus on one or
more of the parties involved in the structure. The First Polar
View and Second Polar View are instances of Polar Views of
the Bond.
   (In contrast with Axial Views.)

Praxeology. 3.11.
   'Praxeology' is another name for Behaviorology, that is, study
of the Behavioral Mode.
   (In contrast with Ethology, Biology, and Physics.)

Preparation. 3.21, 3.26 (3.24).

The Preparation Period of redemption is the OT period. The Adamic Preparation Period is the period of creation (Gen. 1:1–2:3).

(In contrast with Accomplishment and Application Periods.)

Presbyterial. 3.1212.

The Presbyterial Function is that part of the Personal Mode covered by the Kingly and Sabbatical Functions; that is, the Presbyterial Function consists of the bundle of characteristics associated *both* with kingly rule and power *and* with sabbatical worship.

(In contrast with Dogmatical, Diaconal, Juridical, and Technical Functions.)

Presbyteriology. 3.1212.

Presbyteriology is the study of the Presbyterial Function.

(In contrast with Dogmatics, Diaconology, Jurisprudence, and Technology.)

presupposition. (1.1).

A presupposition is a belief or disposition to which one clings for life and death, and which one does not allow to be refuted or overthrown by evidence.

Priestly. 3.1212.

The Priestly or Hieratic Function is that part of the Personal Mode having to do with activities, states, characteristics, etc., of a predominantly priestly sort; that is, the Priestly Function has to do with communion, with sharing in value (blessing and cursing).

(In contrast with Prophetic and Kingly Function.)

Prophetic. 3.1212.

The Prophetic Function is that part of the Personal Mode having to do with activities, states, characteristics, etc., of a predominantly prophetic sort; that is, the Prophetic Function has to do with meaning, communication, wisdom, and information.

(In contrast with Kingly and Priestly Functions.)

Prophetics. 3.1212.

Prophetics is the study of the Prophetic Function.
(In contrast with Basilics and Hieratics.)

Pseudo. 6.120.

Pseudo Study, or Technics, or Beneficence, is such Study, Technics, or Beneficence *against* God and his kingdom.
(In contrast with Genuine Study.)

Quantitative. 10.

The Quantitative Subfunction is that part of the Mathematical Function (more precisely, the Adumbrative Prophetic Function of the Physical Mode) that the Wave View focuses on. That is, it consists of the sequence characteristics found in the "meaning" side of the Physical Mode.
(In contrast with Aggregative and Spatial Subfunctions.)

Reductionism. (3.133).

Reductionism is the attempt to discuss, explain, or account for *some* major categories of Items in terms of *others*. Reductionism occurs as Emphasizing Reductionism (emphasis on some given categories), Exclusive Reductionism (insistence on the innate superiority of some given categories), and Slippery Reductionism (ambiguous use of categories to produce idolatrous Pseudo explanation).
(In contrast with focus and knowledge.)

Refined. 5.32 (6.121).

Refined knowledge or Study is knowledge or Study adapted for communication to men. This generally involves (a) generality of scope, (b) development of method, and (c) attention to justification of results.
(In contrast with Sensitive and Sapiential knowledge.)

Relational. 3.333.

A Relational Unit is a structure that we regard normally in terms of relations among things, enduring more or less through time.

(In contrast with Particulate and Undulatory Units.)

Relationship. 3.333.

A Relationship is a Relational Unit.
(In contrast with Things and Transactions.)

Sabbatical. 3.1211.

The Sabbatical Function consists of that part of the Personal Mode having to do with activities normally characteristic of the sabbath ordinance, or equivalently, of a person's direct relation to God.
(In contrast with Social and Laboratorial Functions.)

Sanctional. 3.321 (3.322).

The Sanctional aspect of a relation is that part of the relation having to do with the Priestly Function. In particular, the Sanctional aspect of the Covenantal Bond is the Covenantal sanctions.
(In contrast with Locutionary and Administrative aspects.)

Sapiential. 5:32 (6.32).

Sapiential knowledge or Study is knowledge or Study involving some of the relations of what one knows to God.
(In contrast with Sensitive and Refined knowledge.)

Science. 6.121.

Science is Natural Science and Social Science. That is, Science is Refined Modal Study by Cosmic Men, when such Study has methodological similarity to or actually *is* Study of the Behavioral, Biotic, and Physical Modes.
(In contrast with Philosophy and Theology.)

Second Polar View. 3.321, 3.322 (3.332, 3.323, 3.3243).

The Second Polar View of the Bond focuses on the role of Creation with respect to the Bond.
(In contrast with First Polar and Axial Views.)

Semidifferentiated. 3.333.

A Semidifferentiated Unit is a Unit which is not Differentiated but which is Weighted in one of the Functions Sabbatical, Social, Laboratorial, Prophetic, Kingly, or Priestly. That is, it is Weighted in one of these six Functions without being discernibly Weighted in one of the Functions that further subdivide the Weighted Function.

(In contrast with Differentiated and Undifferentiated Units.)

Sensitive. 5.32.

Sensitive knowledge is knowledge specialized with respect to subject-matter.

(In contrast with Refined and Sapiential Knowledge.)

Servient. 3.322, 3.332 (3.32433).

A Servient Bond is that part of the Covenantal Bond that pertains to a given Creature.

(In contrast with Dominical and Covenantal Bond.)

A Servient View of a structure is the view from the standpoint of a Servient Bond.

(In contrast with Dominical and Covenantal Views.)

Set Theory. 10.

Set Theory is the study of the Aggregative Subfunction.

(In contrast with Arithmetic and Geometry.)

Situational. 4.1, 4.2 (3.35).

The Situational Perspective on Axiology is that way of looking at the value of Items which focuses on the situation in which the Items occur.

(In contrast with Normative and Existential Perspectives.)

Slippery Reductionism. 3.133.

Slippery Reductionism is the ambiguous use of key terms in a broad sense and in a narrow sense, in order to construct a non-Christian "ultimate explanation" of the Cosmos.

(In contrast with Emphasizing Reductionism and Exclusive Reductionism.)

Social. 3.1211.

The Social Function consists of that part of the Personal Mode having to do with activities normally characteristic of the ordinance of family, or equivalently, of a person's relation to the Cosmic Human Kingdom.

(In contrast with Sabbatical and Laboratorial Functions.)

Social Science. 6.121.

Social Science is Refined Modal Study by Cosmic Men, of the Personal Mode and various Functions within it, especially when such Study has methodological similarity to Natural Science. (In contrast with Natural Science and Humanities.)

Societal Unit. 3.333.

A Societal Unit is a Unit including Men in its internal substructure.

Sociology. 3.1211.

Sociology is the study of the Social Function. (In contrast with Liturgiology and Ergology.)

Spatial. 10.

The Spatial Subfunction is that part of the Mathematical Function (more precisely, of the Adumbrative Prophetic Function of the Physical Mode) that the Field View focuses on. That is, it consists of the interdependent characteristics and relations found in the "meaning" side of the Physical Mode. (In contrast with Aggregative and Quantitative Subfunctions.)

Special. 6.122.

Special Study is Study focused on some agreed upon subject-matter with some agreed upon methodology and justification. The agreement takes place within a group of Students that may be large or small. If the agreement is more or less explicit or conscious or well worked out, we have a case of Special Refined Study. (In contrast with Boundary Study.)

Speculative. 6.123.

Speculative Study is the Study of the Dominical Word of God (especially as this goes beyond the Covenantal Word of God), with the purpose of communicating what this Word says. (In contrast with Canonical and Evangelical Study.)

Strategic. 3.333.

Strategic Institutions are (normally non-Obligatory) Institutions which, in many situations, people with particular callings are virtually obliged to join in order to fulfill those callings.

(In contrast with Obligatory and Voluntary Institutions.)

Structural Study. (6.121).

Structural Study is Study whose subject-matter is (chiefly or focally) structurality.

(In contrast with Modal and Temporal Study.)

structurality. (3.3, 3.35).

Structurality is the interdependent characteristics and relations of Items, especially of Kingdoms, of Creatures, and of God.

(In contrast with modality and temporality.)

Study. 6.120 (6.).

Study is Personal activity with Prophetic Weight, or the result of such activity.

(In contrast with Technics and Beneficence.)

Subfunction. 10.

Subfunctions are subdivisions within or parts of Functions, obtained by considering Functions in terms of several Views. The only Subfunctions explicitly defined in this book are the Aggregative, Quantitative, and Spatial Subfunctions.

(In contrast with Modes and Functions.)

Subhuman Kingdom. 2.40 (2.43).

The Subhuman Kingdom is that part of nonhuman Creation placed in Genesis 1 under man's dominion and rule.

(In contrast with the Human Kingdom and Heaven.)

Systematic Theology. 6.123.

Systematics Theology is Theology that answers the question, "What does the Bible as a whole say?"

(In contrast with Exegesis and biblical theology.)

Technical. 3.1212.

The Technical Function is that part of the Personal Mode covered by the Kingly and Laboratorial Functions; that is, the Technical Function consists of the bundle of characteristics

associated *both* with kingly rule and power *and* with man's labor.
(In contrast with Cognitional, Aesthetic, Presbyterial, and
Juridical Functions.)

Technology. 3.1212.
> Technology is the study of the Technical Function.
> (In contrast with Logic, Aesthetics, Presbyteriology, and
> Jurisprudence.)

Technics. 6.120.
> Technics is Personal activity with Kingly Weight, or the result
> of such activity.
> (In contrast with Study and Beneficence.)

Temporal Study. 6.121.
> Temporal Study is Study whose subject-matter is (chiefly or
> focally) temporality.
> (In contrast with Modal and Structural Study.)

temporality. (3.2, 3.35), (3.0).
> Temporality is the developmental and sequence characteristics
> of Items, especially of Kingdoms, of Creatures, and of God.
> (In contrast with modality and structurality.)

Terrestrial Kingdom. 2.431 (2.432).
> The Terrestrial Kingdom is that part of the Cosmos consisting
> of the land and its inhabitants.
> (In contrast with the Aquatic Kingdom and Heaven.)

Theology. 6.123.
> Theology is Refined Evangelical Study by Cosmic Men.
> (In contrast with Science and Philosophy.)

Theology Proper. 2.1.
> Theology Proper is the study of God.
> (In contrast with Ktismatology and Incarnate Christology.)

Theriology. 2.40.
> Theriology is the study of the Subhuman Kingdom.
> (In contrast with Anthropology and Ouranology.)

Thing. 3.333.

A Thing is a Particulate Unit.

(In contrast with Transactions and Relationships.)

Transaction. 3.333.

A Transaction is an Undulatory Unit.

(In contrast with Things and Relationships.)

Undifferentiated. 3.333.

An Undifferentiated Unit is a Unit not discernibly Weighted in one Function rather than in another.

(In contrast with Differentiated and Semidifferentiated Units.)

Undulatory. 3.333.

An Undulatory Unit is a structure that we regard normally in terms of process, as a unified whole of events.

(In contrast with Particulate and Relational Units.)

Unit. 3.333.

A Unit is a Particulate, Undulatory, or Relational Unit. It is a composite whole, recognizable by observers within a system of composite wholes. It has a certain amount of variation but is in contrast with other wholes. A Unit has been well described when there have been specified its contrastive-identificational features, its variation, and its distribution.

(In contrast with Item and Creature.)

Variation. 5.21.

The Variation of an Item is the range of difference through which it may vary while still remaining recognizably the "same."

(In contrast with Contrast and Distribution.)

View. 3.123, 3.25, 3.321, 3.322, 3.332, 3.35.

A View is a way of looking at Items with a certain perspective or focus. The three main Views are the Particle View, the Wave View, and the Field View. Other Views are Generational, Developmental, and Culminational Views; First Polar, Axial, and Second Polar Views of the Bond; Polar and Axial Views; Dominical, Covenantal, and Servient Views.

(In contrast with Item and Man. See Perspective.)

Vocative. 3.27.
The Vocative aspect of events is that aspect involving the Prophetic Function. Sometimes there can be a temporal separation between a Vocative phase and Dynamic and Appraisive phases.
(In contrast with Dynamic and Appraisive aspects.)

Voluntary. 3.333.
Voluntary Institutions are those in which membership is normally determined by personal considerations, not tightly bound up with a man's major calling.
(In contrast with Obligatory and Strategic Institutions.)

Wave. 3.123 (3.35).
A Wave View is a view focusing on sequence characteristics of Items, not requiring sharp segmentation at the borders of Items.
(In contrast with Particle and Field Views.)

Weight. 3.333.
A Unit is Weighted in X or has X Weight when the X Function or Mode stands in prominence in the Unit's characteristics.

Word of God. 3.3242.
The Word of God is the Dominical Locution of God. Or, less technically, the Word of God is what God says.
(In contrast with Administration and Sanctions of God. *See* Law.)

Zoology. 2.432.
Zoology is the study of the Animal Kingdom.
(In contrast with Botany and Inorganics.)

BIBLIOGRAPHY

Allis, Oswald T. *The Five Books of Moses.* Philadelphia: Presbyterian and Reformed, 1969.

Barr, James. *The Semantics of Biblical Language.* London: Oxford, 1961.

Bartsch, Hans Werner, ed. *Kerygma and Myth; a Theological Debate.* London: S.P.C.K., 1957.

Bavinck, Herman. *The Philosophy of Revelation . . . .* New York: Longmans Green, 1909.

Berkouwer, Gerrit C. "Christian Faith and Science." *Free University Quarterly* 4 (1955–57), pp. 3-10.

―――. *The Providence of God.* Grand Rapids: Eerdmans, 1961.

The Bible. *New American Standard Bible.* Carol Stream, Ill.: Creation House, 1971.

―――. *The New Berkeley Version.* Rev. ed. Grand Rapids: Zondervan, 1969.

―――. *The New English Bible.* Oxford: Oxford University, 1970.

―――. *Revised Standard Version.* Camden, N. J.: Nelson, 1952.

Bohr, Niels. *Atomic Physics and Human Knowledge.* New York: Science Editions, 1961.

Bube, Richard, ed. *The Encounter Between Christianity and Science.* Grand Rapids: Eerdmans, 1968.

―――. *The Human Quest: a New Look at Science and Christian Faith.* Waco, Tex.: Word, 1971.

Bultmann, Rudolf. *Theology of the New Testament.* Vol. I. New York: Charles Scribner's Sons, 1951.

Burtt, Edwin Arthur. *The Metaphysical Foundations of Modern Physical Science.* Rev. ed. Garden City, N. Y.: Doubleday, 1954.

Butterfield, Herbert. *The Origins of Modern Science; 1300–1800.* Rev. ed. New York: Free Press, 1965.

Buttrick, George Arthur, ed. *The Interpreter's Dictionary of the Bible.* 4 vols. New York: Abingdon, 1962.

Calvin, John. *Commentaries on the First Book of Moses Called Genesis.* Grand Rapids: Eerdmans, 1948.

———. *Commentary on the Epistles of Paul the Apostle to the Corinthians.* Edinburgh: Calvin Translation Society, 1848.

———. *Institutes of the Christian Religion.* Grand Rapids: Eerdmans, 1964.

Carnell, Edward John. *A Philosophy of the Christian Religion.* Grand Rapids: Eerdmans, 1952.

Clark, Gordon H. *A Christian Philosophy of Education.* Grand Rapids: Eerdmans, 1946.

———. *A Christian View of Men and Things.* Grand Rapids: Eerdmans, 1952.

———. *The Philosophy of Science and Belief in God.* Nutley, N. J.: Craig, 1964.

———. *Religion, Reason and Revelation.* Philadelphia: Presbyterian and Reformed, 1961.

Clowney, Edmund P. *The Biblical Doctrine of the Church.* Unpublished. Philadelphia: Westminster Theological Seminary, c. 1968. (Mimeographed.)

The Confession of Faith . . . . Edinburgh: Free Presbyterian Church of Scotland, 1967.

Diepenhorst, I. A. "Science, Its Nature, Its Possibilities, and Its Limitations." *Free University Quarterly* 4 (1955-57), pp. 11-37.

Dooyeweerd, Herman. *In the Twilight of Western Thought.* Nutley, N. J.: Craig, 1965.

———. *A New Critique of Theoretical Thought.* 4 vols. Reprint. Philadelphia: Presbyterian and Reformed, 1969.

———. *The Secularization of Science.* n.d. (Mimeographed.)

Douglas, J. D. *The New Bible Dictionary.* London: Inter-Varsity, 1962.

Driver, Samuel R. *The Book of Genesis.* 11th ed. London: Methuen, 1920.

Eddington, Arthur. *Space, Time, and Gravitation* . . . . Cambridge: Cambridge University, 1921.

Feigl, Herbert, and May Brodbeck, eds. *Readings in the Philosophy of Science.* New York: Appleton-Century-Crofts, 1953.

Feigl, Herbert, and Wilfrid Sellars, eds. *Readings in Philosophical Analysis.* New York: Appleton-Century-Crofts, 1949.

Fletcher, Joseph. *Situation Ethics; the New Morality.* Philadelphia: Westminster, 1966.

Frame, John M. *The Amsterdam Philosophy; a Preliminary Critique.* Phillipsburg, N. J.: Harmony Press, *c.* 1972.

————. "What Is God's Word." *Presbyterian Guardian* 42 (November, 1973), pp. 142-143.

————. "The Word of God and the AACS; a Reply to Professor Zylstra." *Presbyterian Guardian* 42 (April, 1973), pp. 60-61.

Geehan, E. R., ed. *Jerusalem and Athens.* Philadelphia: Presbyterian and Reformed, 1971.

Guthrie, Donald, *et al.,* eds. *The New Bible Commentary Revised.* Grand Rapids: Eerdmans, 1970.

Harrison, R. K. *Introduction to the Old Testament.* Grand Rapids: Eerdmans, 1969.

Hart, Hendrik. "Problems of Time: an Essay." *Philosophia Reformata* 38 (1973), pp. 30-42.

Heisenberg, Werner. *Physics and Philosophy; the Revolution in Modern Science.* New York: Harper & Brothers, 1958.

Hempel, Carl G. *Fundamentals of Concept Formation in Empirical Science.* International Encyclopedia of Unified Science, vol. II, no. 7. Chicago: University of Chicago, 1952.

Hillers, Delbert R. *Treaty-Curses and the Old Testament Prophets.* Rome: Pontifical Biblical Institute, 1964.

Hodge, Charles. *A Commentary on the First Epistle to the Corinthians.* Reprint. London: Banner of Truth Trust, 1964.

Hooykaas, R. *Natural Law and Divine Miracle; the Principle of Uniformity In Geology, Biology and Theology.* 2nd impression. Leiden: Brill, 1963.

————. *Philosophia Libera; Christian Faith and the Freedom of Science.* London: Tyndale, 1957.

Jackson, Samuel, ed. *The New Schaff-Herzog Encyclopedia of Religious Knowledge.* 13 vols. Grand Rapids: Baker, 1951-53.

Jeans, James. *Physics & Philosophy.* Cambridge: Cambridge University, 1943.

Jeeves, Malcolm. *The Scientific Enterprise and Christian Faith.* Downers Grove, Ill.: Inter-Varsity, 1969.

Judge, Joseph. "The Zulus: Black Nation in a Land of Apartheid. *National Geographic* 140 (December, 1971), pp. 738-775.

Kitchen, K. A. *Ancient Orient and the Old Testament.* Chicago: Inter-Varsity, 1966.

Kittel, Gerhard, and Gerhard Friedrich, eds. *Theological Dictionary of the New Testament.* 9 vols. Grand Rapids: Eerdmans, 1964-74.

Kline, Meredith G. *By Oath Consigned.* Grand Rapids: Eerdmans, 1968.

————. *The Structure of Biblical Authority.* Grand Rapids: Eerdmans, 1972.

————. *Treaty of the Great King; the Covenant Structure of Deuteronomy Studies and Commentary.* Grand Rapids: Eerdmans, 1963.

Kuhn, Thomas. *The Structure of Scientific Revolutions.* International Encyclopedia of Unified Science. Vol. II, no. 2. 2nd ed. Chicago: University of Chicago, 1970.

Kuyper, Abraham. *Encyclopedia of Sacred Theology.* London: Hodder and Stoughton, 1899.

————. *The Work of the Holy Spirit.* Grand Rapids: Eerdmans, 1941.

Lee, Francis Nigel. *A Christian Introduction to the History of Philosophy.* Nutley, N. J.: Craig, 1969.

————. *The Covenantal Sabbath.* London: Lord's Day Observance Society, c. 1973.

————. *Calvin on the Sciences.* Cambridge: Sovereign Grace Union, 1969.

Leith, T. H. "Notes on the Predispositions of Scientific Thought and Practice." *Journal of the American Scientific Affiliation* 24 (June, 1972), pp. 51-53.

Lewis, Clive S. *Miracles; a Preliminary Study.* London: Centenary, 1947.

Maatman, Russell. *The Bible, Natural Science, and Evolution.* Grand Rapids: Baker, 1970.

————. "Can the Bible Contain Scientific Facts?" *Torch and Trumpet* 20 (February, 1970), pp. 14-16.

McCarthy, Dennis J. *Treaty and Covenant.* Rome: Pontifical Biblical Institute, 1963.

McClain, Alva J. *The Greatness of the Kingdom . . . .* Chicago: Moody, 1959.

Madden, Edward H. "Scientific Explanations." *Review of Metaphysics* 26 (June, 1973), pp. 723-743.

Mendenhall, George E. *Law and Covenant in Israel and the Ancient Near East.* Pittsburgh: Biblical Colloquium, 1955.

Montgomery, John W. *The Shape of the Past; an Introduction to Philosophical Historiography.* Ann Arbor: Edwards Brothers, 1962.

————. *Where Is History Going? Essays in Support of the Historical Truth of the Christian Revelation.* Grand Rapids: Zondervan, 1969.

Morris, Henry. *Biblical Cosmology and Modern Science.* Nutley, N. J.: Craig, 1970.

Murray, John. *Divorce.* Philadelphia: Committee on Christian Education, Orthodox Presbyterian Church, 1953.

————. *The Imputation of Adam's Sin.* Grand Rapids: Eerdmans, 1959.

————. *Principles of Conduct; Aspects of Biblical Ethics.* Grand Rapids: Eerdmans, 1957.

Nash, Ronald H., ed. *The Philosophy of Gordon H. Clark.* Philadelphia: Presbyterian and Reformed, 1968.

North, Gary, and Rousas Rushdoony, eds. *Foundations of Christian Scholarship: Essays in the Van Til Perspective.* To appear.

Orr, James, ed. *International Standard Bible Encyclopedia.* 5 vols. Grand Rapids: Eerdmans, 1939.

Payne, J. Barton, ed. *New Perspectives on the Old Testament.* Waco, Tex.: Word, 1970.

Pepper, Stephen C. *World Hypotheses; a Study in Evidence.* Berkeley: University of California, 1970.

Philosophy and Christianity; Philosophical Essays Dedicated to Professor Dr. Herman Dooyeweerd. Kampen: Kok, 1965.

Piel, Gerard. *Science in the Cause of Man.* New York: Knopf, 1961.

Pike, Kenneth L. "Foundations of Tagmemics—Postulates—Set I." Unpublished. 1971.

————. "Language as Particle, Wave, and Field." *The Texas Quarterly* 2 (Summer, 1959), pp. 37-54.

————. *Language in Relation to a Unified Theory of the Structure of Human Behavior.* 2nd ed. The Hague-Paris: Mouton, 1967.

————. *Linguistic Concepts.* Unpublished. 1968.

————. *Selected Writings to Commemorate the 60th Birthday of Kenneth Lee Pike,* ed. Ruth M. Brend. The Hague-Paris: Mouton, 1972.

Polanyi, Michael. *Personal Knowledge; Towards a Post-Critical Philosophy.* Chicago: University of Chicago, 1958.

————. *Science, Faith and Society.* Chicago: University of Chicago, 1964.

————. *The Tacit Dimension.* London: Routledge & Kegan Paul, 1967.

Popper, Karl R. *The Logic of Scientific Discovery.* New York: Basic Books, 1959.

Ramm, Bernard. *The Christian View of Science and Scripture.* Grand Rapids: Eerdmans, 1954.

Ream, Robert. *Science Teaching: A Christian Approach* Philadelphia: Presbyterian and Reformed, 1972.

Reichenbach, Hans. *The Philosophy of Space and Time.* New York: Dover, 1958.

Reid, James. *God, the Atom, and the Universe.* Grand Rapids: Zondervan, 1968.

Reymond, Robert: *A Christian View of Modern Science.* Philadelphia: Presbyterian and Reformed, 1964.

Robertson, Archibald, and Alfred Plummer. *A Critical and Exegetical Commentary on the First Epistle of St. Paul to the Corinthians.* Edinburgh: T. & T. Clark, 1911.

Runner, H. Evan. *Syllabus for Philosophy 220; the History of Ancient Philosophy.* Unpublished. Grand Rapids: Calvin College, 1958–1959.

Rushdoony, Rousas J. *The Institutes of Biblical Law.* Nutley, N. J.: Craig, 1973.

————. *The Mythology of Science.* Nutley, N. J.: Craig, 1967.

Rust, Eric C. *Science and Faith; Towards a Theological Understanding of Nature.* New York: Oxford, 1967.

Ryrie, Charles C. *Dispensationalism Today.* Chicago: Moody, 1965.

Schaff, Philip. *The Creeds of Christendom, with a History and Critical Notes.* 3 vols. Grand Rapids: Baker, 1966.

Schiaparelli, Giovanni. *Astronomy in the Old Testament.* Oxford: Clarendon, 1905.

Schilpp, Paul Arthur, ed. *Albert Einstein: Philosopher-Scientist.* Evanston, Ill.: The Library of Living Philosophers, 1949.

————. *The Philosophy of Rudolf Carnap.* LaSalle, Ill.: Open Court, c. 1963.

Seerveld, Calvin. *A Christian Critique of Literature . . . .* Christian Perspectives Series 1964. Hamilton, Ontario: Association for Reformed Scientific Studies, c. 1964.

Shepherd, Norman. "The Doctrine of Scripture in the Dooyeweerdian Philosophy of the Cosmonomic Idea." *Christian Reformed Outlook* 21 (February, 1971), pp. 18-21; (March, 1971), pp. 20-23.

Smart, John J. C., ed. *Problems of Space and Time.* New York: Macmillan, 1964.

Spier, J. M. *An Introduction to Christian Philosophy.* Philadelphia: Presbyterian and Reformed, 1954.

————. *Tijd en eeuwigheid* . . . . Kampen: Kok, 1953.

Steen, Peter. *The Idea of Religious Transcendence in the Philosophy of Herman Dooyeweerd.* Unpublished. Chestnut Hill, Pa.: Westminster Theological Seminary, Ph.D. thesis, 1970.

————. *The Supra-temporal Selfhood in the Philosophy of Herman Dooyeweerd.* Unpublished. Chestnut Hill, Pa.: Westminster Theological Seminary, Th.M. thesis, 1961.

Stoker, Hendrik G. *Beginsels en metodes in die wetenskap.* Potchefstroom: Pro-Rege-Pers, 1961.

————. *Oorsprong en Rigting.* 2 vols. Kaapstad: Tafelberg, 1967, 1970.

————. *Die wysbegeerte van die skeppingsidee* . . . . Pretoria: de Bussy, 1933.

Strauss, Daniël F. M. *Wysbegeerte en vakwetenskap.* Bloemfontein: Sacum, c. 1973.

Torrance, Thomas F. *Space, Time, and Incarnation.* London: Oxford University, 1969.

————. *Theological Science.* London: Oxford University, 1969.

Toulmin, Stephen E. *The Uses of Argument.* Cambridge: Cambridge University, 1969.

van der Laan, H. *A Christian Appreciation of Physical Science* . . . . Hamilton, Ontario: Association for Reformed Scientific Studies, c. 1967.

van der Merwe, Nicolaas T. *Op weg na 'n christelike logika.* Potchefstroom: University of Potchefstroom, Ph.D. thesis, 1958.

van Riessen, Hendrik. *The Christian Approach to Science.* Association for the Advancement of Christian Scholarship, n.d. (Mimeographed.)

————. *Wijsbegeerte.* Kampen: Kok, 1970.

Van Til, Cornelius. In Defense of the Faith, vol. III. *Christian Theistic Ethics.* Philadelphia: Presbyterian and Reformed, 1971.

————. In Defense of the Faith, vol. VI. *Christian-theistic Evidences.* Philadelphia: Presbyterian and Reformed, 1961.

————. *A Christian Theory of Knowledge*. Philadelphia: Presbyterian and Reformed, 1969.

————. *The Defense of the Faith*. Philadelphia: Presbyterian and Reformed, 1955.

————. *The Defense of the Faith*. 3rd rev. ed. Philadelphia: Presbyterian and Reformed, 1967.

————. *The Dilemma of Education*. National Union of Christian Schools, 1954.

————. In Defense of the Faith, vol. IV. *An Introduction to Systematic Theology*. Philadelphia: Presbyterian and Reformed, 1966.

————. In Defense of the Faith, vol. II. *A Survey of Christian Epistemology*. Philadelphia: Presbyterian and Reformed, 1970.

Vollenhoven, Dirk H. Th. "Hoofdlijnen der logica." *Philosophia Reformata* 13 (1948), pp. 59-118.

————. *De noodzakelijkheid eener christelijke logica*. Amsterdam: H. J. Paris, 1932.

Warfield, Benjamin B. *Selected Shorter Writings of Benjamin B. Warfield*—I, ed. John E. Meeter. Nutley, N. J.: Presbyterian and Reformed, 1970.

Weeks, Noel. *Creation Themes in the Psalms*. Unpublished. Chestnut Hill, Pa.: Westminster Theological Seminary, Th.M. thesis, 1968.

Whitcomb, John C., Jr., and Henry M. Morris. *The Genesis Flood . . . .* Philadelphia: Presbyterian and Reformed, 1961.

Whitehead, Alfred North. *Process and Reality, an Essay in Cosmology*. Cambridge: Cambridge University, 1929.

————. *Science and the Modern World*. New York: New American Library, 1949.

Wittgenstein, Ludwig. *Tractatus Logico-Philosophicus*. London: Routledge & Kegan Paul, 1951.

Young, Edward J. *Studies in Genesis One*. Philadelphia: Presbyterian and Reformed, 1964.

Ziff, Paul. *Semantic Analysis*. Ithaca, N. Y.: Cornell University, 1960.